10 Questions Science Can't Answer (Yet)

10 Questions Science Can't Answer (Yet)

A Guide to the Scientific Wilderness

Michael Hanlon

Macmillan

London   New York   Melbourne   Hong Kong

First published 2007 by
Macmillan
Houndmills, Basingstoke, Hampshire RG21 6XS and
175 Fifth Avenue, New York, N. Y. 10010
Companies and representatives throughout the world

ISBN-13: 978-0-230-51758-5 hardback
ISBN-10: 0-230-51758-7 hardback

This book is printed on paper suitable for recycling and made from fully managed
and sustained forest sources.

A catalogue record for this book is available from the British Library.

A catalog record for this book is available from the Library of Congress.

10   9   8   7   6   5   4   3   2   1
16   15   14   13   12   11   10   09   08   07

Printed and bound in the United States of America

contents

acknowledgements

At Macmillan my editor Sara Abdulla has been a constant source of inspiration, encouragement and, occasionally, enforcement, as deadlines approach. I must thank all the scientists, writers and press officers who furnished me with quotes, helpful explanations and who trawled through their databases finding papers whose names, publication dates and authors I had totally forgotten – in particular Claire Bowles at *New Scientist*, Ruth Francis at *Nature*, Peter Barratt at The Particle Physics and Astronomy Research Council and all the helpful staff at the American Association for the Advancement of Science.

I must thank my employer, the Daily Mail, for giving me the space and freedom to develop my interest in science in all its facets, and most of all my partner, Elena Seymenliyska, who has put up with yet another slew of weekends and evenings taken up by my finishing another book.

Michael Hanlon
March 2007

introduction

There is nothing new to be discovered in physics now. All that remains is more and more precise measurement

Lord Kelvin, 1900

How we laugh now at those daft Victorians. They thought they knew everything. To them, the Universe was a small and well-ordered sort of place, consisting of a few million stars. The planets were held aloft by Newton's well-ordered apron strings, and the whole cosmos ticked away like a Swiss clock.

Down here on Earth they knew that life began in a warm little pond, and that its subsequent evolution was governed by Mr Darwin's grand thesis. Stuff was made of atoms, of about a hundred different flavours, which behaved like mini versions of the planets: tiny, well-behaved billiard balls. Science was nearing its end – all that was left was to cross the 't's and dot the 'i's. We were nearly at the Summit of Total Understanding.

The summit turned out to be a false one. A whole series of brilliant and bothersome insights in the early 20th century threw so many spanners into the scientific works that we were forced to more or less rip everything up that we thought we knew and to start again.

We now know that the Universe is rather larger, and more ancient, than Kelvin and his contemporaries imagined. We know how the stars shine, what they are made of and how they evolve. We have plotted the age of the Earth and of the planets, and have discovered or inferred some fantastic monsters unknown to the Victorians – the quasars, neutron stars and black holes.

We have confirmed Darwin's theories (no serious paper challenging the basic reality of evolution has been published for
more than a century) and now have a mechanism, genetics,
to explain how information is copied from one generation to
the next. Newton's clockwork, although a brilliantly accurate
description of the parochial, we now know breaks down on
cosmic scales and at the most rapid velocities. Einstein linked
space and time, acceleration and gravity, and mass and
energy in wonderful and counter-intuitive ways.

Atoms, previously indivisible, have been shown to consist of
a bizarre and frankly random-seeming troublesome menagerie of fundamental particles. And on their scale, things
behave differently again from the large, the heavy and the
rapid described by Einstein. It seems that the Universe can be
very odd – spooky even. Particles can influence other particles
billions of light years apart, instantaneously. There may be
many more dimensions than the three of space and one of
time with which we are familiar. Moving clocks run slow.

Science now has theories of how the mind works, and have
gained great insights into human psychology. We have slain
many of the great killers of humankind with our medicine, and
are on the way to slaying more. We have even taken the first
steps towards exploring our nearby Universe. Some of our
technology, in the words of Arthur C. Clarke, is so advanced
that to people a hundred years ago it would be indistinguish-
able from magic.

So what more must there be to know?

A lot is the answer, and that is what this book is about. There
are no great answers here, just some big (and not so big)
questions, things that science has often sidestepped or before
which it has simply wilted: challenges that have been considered too enormous, at least until recently.

Perhaps our hubris is as great, perhaps greater, than that of
the Victorians. That is not because we know less, but perhaps

because we know more. The gargantuan and sustained acceleration in scientific discovery and technological progress in the last century, especially since the end of the Second World War, has led to the widespread belief that the Great War on Ignorance – the final phase of which began in the Enlightenment more than 200 years ago – is within a measurable distance of its end.

After all, we've mapped the human genome and ripped atoms asunder into their constituent parts. A grand theory of everything, uniting the fundamental forces and reconciling the quantum world with relativity, may be only a few years away. We are close to understanding the building blocks of life, and every week it seems that astronomers make a new discovery which revolutionizes our understanding of the Universe. Our technology really does seem magical. Imagine trying to explain the Web to Lord Kelvin.

And yet. Like him, we still have no idea what the Universe is made of (or at least most of it). The Victorians presumed it was made of just atoms, and how wrong they were. The great bulk seems to be made of two mysterious substances, dark matter and dark energy, whose nature we can only guess at. The ordinary stuff (aka *baryonic matter* – good old hydrogen, you and me, Neptune and your dining table) is just a sideshow. Most of everything is invisible and we have no idea at all what it is – just a few rather vague guesses.

We have a theory, or rather a model, for the origin of our Universe, the Big Bang, but we do not really understand what it was that banged, how it banged, or what lit the touchpaper. We do not know what came before the start of our universe, or even whether asking this question makes any sense. We do not even know whether the universe we think of as our Universe really is all there is. Those Victorians were out by several orders of magnitude when they considered the Milky Way to be all there was.

We may well be underestimating the scale of things by dozens more orders of magnitude again when you consider that physics now takes seriously the idea of the multiverse: a vast, indeed infinite, ensemble of cosmoses that may be required to solve one of the great riddles – why our Universe seems to be so finely tuned as to allow for our existence. It could be that what we see in the night sky, all that cosmic glory, the whirlpool galaxies and quasars, the endless voids of intergalactic space, is just a pimple on the backside of something infinitely grander. The true nature of reality is as elusive as ever.

More mysteries: we do not know how life began on our Earth, nor whether it has ever begun elsewhere. We are probably no closer to understanding the true nature of the human mind than Plato or Aristotle. We have no idea how exactly a couple of pounds of grey jelly can come up with *Romeo and Juliet*, appreciate a nice sunset or be in agonizing pain.

The nature of time eludes us, and the ageing process is still something of a puzzle. We still do not know, really, why we get old, nor why we get old at the rate we do, and nor do we know whether there is anything we will be able to do about this.

Much more about humanity remains puzzling. As a species we seem to be changing shape – literally. The obesity epidemic sounds straightforward, but in fact there are some mysteries here. And our politicians and social scientists still resolutely fail to deal with, or even acknowledge, the fundamental differences between individuals that cause so much pain and grief.

The minds of our fellow creatures are still deeply mysterious, although the more we learn, the more we are forced to conclude that the great gap between humans and beasts may not be as wide as we thought. What we know is hugely impressive, but the more one thinks about what we don't know, the more one is forced to conclude that rather than having waded deep into the sea of knowledge we have merely dipped our toes in the water.

The sheer scale of what we don't know still dwarfs what we do, but you perhaps wouldn't think so reading the papers. To read the journals, the newspapers and the magazines, and to watch the TV programmes that have often done a brilliant job of popularizing some of the most difficult concepts, one could be forgiven for thinking that most of science is 'in the bag'.

Science now comes equipped with its own *commentariat*, of press officers, professional spin doctors and Internet sites, and of course a whole army of journalists who specialize in the subject. Today, in the UK, there are perhaps a hundred or so people like me who make their living writing about science for newspapers, the Web and magazines, or making TV and radio programmes. Fifty years ago there were fewer than a dozen.

They are needed because many scientists have themselves often been very bad at talking about what they do. (Isaac Newton, possibly the greatest mind of historic times, was, famously, so terrible at communicating that his lectures were often only attended by one or two bored students.)

Even today many scientists do not possess mobile phones, have no idea how to use them if they do and only check their email once a week. I know one or two who still use faxes. There have always been scientists like this, but the difference is that nowadays they have spin doctors, and as a result science has become something like politics used to be, a vast ocean of shifting opinions, tides of fashion that wash in and out (this week it's climate change, next we'll start worrying about GM foods again. Hang on a minute... we must be long overdue for a flu scare).

All this has made boffins and their works world-class news. A newspaper editor once told me: 'You are lucky to be a science writer... in the old days, science was nothing, it was all crime, politics and trade unions. Now nobody is interested in all that stuff, it's all global warming, cloned sheep and life on Mars these days'. And she was right. Science has become sexy – has made itself sexy – with the willing collaboration of witterers like me. But there is a problem.

All this reportage is not only over-egging, but spoiling the pudding. Whereas great discoveries once took weeks or even months to come to the public's attention (there was no mention in the newspapers of the discovery of the DNA double helix in 1952 for nearly half a year), now every scientific twist and turn is reported as if it is the final word on the matter and that the riddle has been solved. *Nature*, which reported the DNA story, now employs a full-time press staff to decide which 'stories' that week are newsworthy enough to be worth troubling the hacks with.

Medical journalism, in particular, suffers from some nasty side-effects of over-reporting. In decades past, news from clinical trials and large studies examining the effects of things like diet and lifestyle leaked out slowly. Often this meant we were simply ill-informed; many scientists were pretty convinced of the link between smoking and lung cancer long before the publication of Richard Doll's definitive study in 1954 (in fact, the link was first shown by Nazi scientists in the 1930s, something which is still not fully acknowledged, perhaps understandably), and yet the message took a long time to get out that smoking could kill you. These days even the whiff of something as comprehensive as the Doll study would lead to a thousand magazine articles and front-page headlines, and probably calls for tobacco to be banned forthwith.

There was an upside to this glacial spread of information from the scientific to the public domains. Something like an informal process of peer-review was taking place – which meant that only really significant studies, generating really significant (in statistical terms) results got reported at all.

These days, you can give a few dozen people food additive x and find a tiny statistical effect on their health; then ring the right hacks and you will get immediate calls for x to be banned straight away. The next time you read that substance or activity x has been 'linked' to malaise y, be suspicious – very suspicious. As I write this I see the headline on a London newspaper: 'Drinking cola harms bones'. We have no reason to doubt the results of this research, and yet one has no doubt either that in a few months (or weeks or years) along will come another study showing a statistical link between drinking fizzy pop and good health, maybe even good bones. (Coffee seems to go through a healthy–unhealthy cycle with a regular three-month period. Maybe some strange physical law is operating here.) What these headlines do – and the research that generates them – is to add to the general background hum of pseudo-knowledge, the modern equivalent of all those old wives' tales that we sort of never believed yet quoted endlessly.

But while we now laugh at 'advice' to avoid 'indigestible' food, put butter on a burn and avoid masturbation for fear of blindness or insanity, and we pour scorn on the ancient terror of 'chills', we have created a new folklore of our own – 'too much tea is dehydrating' (it isn't), 'all fat is bad for you' (you would die without some fat), mineral water is purer and safer than tap (nonsense) and so on. The 'news' can be good or bad of course. I trawled some recent stories about heart disease, cancer and stroke and, if all these 'studies' were to be believed, I found that by eating and drinking just the right quantities of mangoes, coffee, red wine, broccoli, fish oil and wild rice, and going to live in the right bit of Japan I could expect to see in the 23rd century. Similarly, I could have read another tranche of articles and found that, with the lifestyle I actually *do* enjoy, I should have been dead in 1989.

And this illusion of progress is certainly not confined to health. In physics, the Grand Theory of Everything has been 'about 20 years away' for at least – well, certainly a lot longer than 20 years. Book after book has been published in the last quarter of a century or so stating quite confidently that while we are not quite there yet, this or that new theory will unite the forces, reconcile the menagerie of fundamental particles, and bring and end to the long war between relativity and quantum physics. But – surprise, surprise – after a succession of false summits, the goal of a Grand Theory of Everything looks as far away as ever.

Sometimes the illusion of progress has serious repercussions, political and social. Climate change is not just a scientific issue; it is highly political as well. On the one hand you have a fairly strong consensus that human activity, mostly the burning of fossil fuels, is making a large and damaging contribution to the global warming that has been measured since 1900. Every month, every week, it seems there is a new piece of evidence that seems to confirm this view. And there is little doubt that the consensus is right. We *are* heating up our planet.

But of course it is not quite as straightforward as that. As the entertaining British politician and journalist Boris Johnson has pointed out, global warming is one of those things one *believes* in or does not, and that there is the whiff of faith about the whole climate change debate. Belief and science are supposed to be incompatible. You do not *believe* in gravity, or in Kepler's Laws, nor in DNA; these things just *are* (or, of course, *are not*). If you leap from a balcony your dismissal of Newton's laws will not help you to land without breaking your bones.

Science has never been so glamorous or so productive. It attracts the most brilliant minds of our generation. But there is something missing: the madness of old. Because as well as being brilliant and inspiring, today's science is an industry. Scientists work in teams, their discoveries and results incremental, their livelihoods depending on rapid publication and peer review. It is a good process, and hard to think of a better one, but it leaves little room for the lunatic and the eccentric. The elitism and *ad hoc* nature of natural philosophy led to a lot of blind alleys, but also a lot of inspiration and brilliance. Darwin's epic voyage round the world in the 1830s led to one of the most brilliant and important insights in the entire history of science, yet if you read his diaries you will realize what a shambles, by modern standards, the whole thing was. It took them several weeks to get the *Beagle* out of Plymouth for a start. And then they couldn't land in Tenerife because of quarantine rules. Now, of course, there would be armies of fixers for any sort of expedition. Today, as never before, the 'maverick' has a hard time in science. People who come up with theories that challenge the consensus tend to be ridiculed, often unfairly.

Yet even today there are free thinkers who sometimes get round the dead hand of the consensus, people like Barry Marshall (who discovered that stomach ulcers are caused by an infection, not stress) and Stanley Prusiner, who proved that the diseases scrapie and CJD were caused by a hitherto unknown infectious agent, a protein called a prion. These islands of brilliance punctuate an often rather flat sea of conformity.

The trouble with the big questions – the where-did-life-come-from questions – is that it is very hard indeed to get research grants to try to answer them. A feature of the questions in this book is that they often do not fall neatly into any one discipline; to truly understand time, for instance, one probably needs to be a physicist, a cognitive psychologist and a philosopher.

To answer the problem of life's origins one needs to know a lot about geology, astronomy, possibly cosmology (let's assume we don't need theology) and biochemistry. Announce you want to investigate the paranormal or the nature of reality and you will have a hard time getting a grant at all. Of course, plenty of people *are* working on these sorts of problem, but polymaths are relatively few in number and tend to be on the fringes.

Today, specialism and scientific correctness is all. The scientific mainstream is channelled, focused and intensely specialized, to the extent that these days a biochemist working on one sort of protein structure is likely to know little of the latest research on, say, fats or carbohydrates, let alone anything much outside of biochemistry.

In the 1840s it was just about possible for Charles Darwin to have a working knowledge of the cutting edge of geology, biology and meteorology and to master the various sub-disciplines within them. Nowadays there is simply so *much* complexity – and jargon – that a Darwin simply could not emerge. And if he did he would never get a research grant to set sail on the *Beagle*.

There are mysteries because a lot of stuff is hard. There are also mysteries that may turn out to be not so mysterious. Science can waste of lot of time chasing rainbows. Take the 'consciousness problem'. From being studiously ignored for decades, what it is that makes minds self-aware is now occupying some of the best brains on the planet. Consciousness is deeply sexy. But not everyone is convinced that the feeling of self-awareness, ostensibly so mysterious and significant, is important at all. Indeed, I know one eminent scientist who refuses to discuss consciousness at all at the dinner table. Such talk, he says, 'inevitably leads just to a lot of silly speculation'. Maybe, just maybe, what we call 'consciousness' doesn't really exist...

One of the mysteries in this book is that of time. Time is something we all take for granted, but it is a very slippery beast indeed when you try to define it. Thinking of space and time together as a sort of fabric which knits the Universe together is very useful way of approaching things on the mathematical level. But does it tell us very much about what time actually *is*? Might time in fact be made to disappear, in a puff of logic, by cancelling out all those little *t*s from the equations?

Probably not, but there is a long history of scientists studying things that do not exist. In the 19th century, psychologists in the US studied a condition that was causing widespread concern, particularly in the south of the country. This illness was dubbed *drapetomania*, and its effects were pernicious and financially damaging. Drapetomania was, you see, the 'uncontrollable urge of a negro to escape from slavery'. Slaves, as well as suffering from this irrational lunacy, could also be blighted with *Dysaethesia aethiopica*, or disobedience.

Drapetomania was first diagnosed by a Louisiana doctor called Samuel Cartwright, who managed to get his new disease written up in the *New Orleans Medical and Surgical Journal*. His preferred cure for this and dysaethesia was simple: a sound flogging. How we squirm, now, at this ridiculous example of legitimized racism, but can we be sure that the phenomena we take for granted and argue over today are any more real than drapetomania? What else in the scientific canon might be no more than a cultural artefact, a reflection of the thinking of the times?

When you think about, it the answer to this question is 'lots'. Not just time and consciousness. Not everyone is convinced that 'dark matter' and 'dark energy' are real things. Maybe the equations are wrong, or we are making misobservations and are thus on a Snark hunt.

It is not entirely clear to what degree the elegant world described by the quantum physicists represents reality, or conversely, a mathematical model of it. Modern physics is full of wonderful models and descriptions of reality, and this is a problem. Space–time twisting and stretching like rubber. Dark energy stretching the fabric of reality itself apart. Tiny strings vibrating away, gravitational wormholes and clashing branes giving rise to whole new universes. How many of these things are real, like Newton's apple was real, and how many are simply abstractions, impossible to envisage except through the lens of equation and proof? The strange world of modern physics in particular is now so surreal, so counterintuitive, that it has sadly left the ordinary man and woman far behind (compare this to the works of Darwin and even Newton, which were comprehensible, at least in their basic outline, to millions of educated people at the time).

This book concentrates on ten questions. It is not a comprehensive list of the greatest unanswered questions in science; instead, it is a snapshot, taken at a particular time, of things that have puzzled this author more than most. Some of the questions are obvious, such as the nature of time and the mystery of reality. Others are there simply because, well, I find them fascinating. The obesity epidemic may not tell us anything very fundamental about the nature of the Universe, but it does tell us an awful lot of fascinating things about life on Earth today and our obsessions.

Some of the great questions are not here. The riddle of consciousness has been done to death, and I am inclined to agree with the eminent British biologist mentioned above. Nevertheless, it is a hugely interesting topic. I have addressed it, a little obliquely, in the chapters dealing with animal sentience and the continuity of existence.

Modern physics is truly a nightmare. Perhaps its greatest, and to date completely unanswered, mystery is 'What exactly is going on in quantum physics?'. Paul Dirac's famous answer to this question was 'shut up and calculate!', but this will of course not do.

For instance: when two electrons, separated by huge distance, are 'entangled', meaning that what happens to one has an instantaneous effect on the other, what exactly are we seeing? How is the information getting there faster than light (which is of course banned)? One interpretation is that somehow a 'message' is being sent back in time. Another is that the particles are 'communicating' with reference to a 'universal wavefunction' that extends everywhere.

The role of the conscious mind is deeply odd. How does the act of observing affect what it is being observed, as some interpretations of quantum physics insist? And one 'interpretation' states that every time a quantum event occurs, a whole new universe is created to allow for all the statistically possible outcomes. Fine, but where do all these universes come from?

What else isn't here? I look at the possibility that the Universe is home to myriad life forms, but have left the whole UFO/ET conundrum to others. The existence of life seems to me to be the profound mystery. The existence of intelligent life is perhaps icing on the cake, although it would be an extremely interesting cake if we could find it. In any case, we may not have to look far to find intelligent life; the more scientists probe the inner workings of the animal brain, the more our cousins seem to climb up the intellectual pole.

We still don't know some of the most basic things about ourselves. The purpose of sleep (and dreaming) remains a mystery, although new theories come along every couple of years. Modern medicine is a triumph, and yet the embarrassing truth is that we don't really understand how a lot of it works. Our brains are still deeply mysterious and not just because they generate conscious experience.

We don't know how, or where, memories are stored. And we don't know whether free will is an illusion. Proving that it is and making everyone realize that it is would be one of the greatest affronts to human dignity since the realization that we weren't designed in any god's image, but that doesn't mean that the latter isn't almost certainly true.

There are an awful lot of things out there that we don't know. Here, for you, are just ten of them, but there are hundreds more. It is the job of science to find the answers, and one has no doubt that it will.

The only trouble is, one also has no doubt that, when this little lot is cleared up, the summit of knowledge we will find ourselves upon will be just as false as that on which stood Lord Kelvin, and the peaks of the unknown just as high and just as distant as they were when we mistakenly thought the end of the climb was just one more heave away.

1

is fido a zombie?

A few years ago, I was lucky enough to find myself 11,000 feet up on the side of a volcano in Rwanda. This central African country is really one of the most charming, eccentric and surreal places on Earth. I don't think I have ever been anywhere, certainly not in Africa, that felt more serene, cheerful and at ease with itself. And yet, just a little more than a decade before my visit, this country was consumed by a carnal spasm of bloodlust rarely equalled anywhere in history.

Rwanda is not just a horror story. It also contains some of the oddest and most picturesque scenery on the planet – the 'land of a thousand hills'. Rwanda is also home (along with some of its neighbours) to one of the world's most magnificent animals: the fabulous, critically endangered mountain gorilla.

And it was the gorillas we were here for. I was writing a story for my newspaper about how these huge beasts had managed to cope with decades of civil war and strife in their homelands. There were tales of how poachers were killing and eating the last of these magnificent animals, whose numbers were down to the mid-hundreds. From what I have heard, it seems probable that the great mountain gorilla may well be on its way out.

Mountain gorillas, like their lowland gorilla cousins, chimpanzees, bonobos, the orang-utans of the East Indies and, of course, us, comprise the great apes. We hold our heads high as if atop the evolutionary tree, although we deserve no such accolade. All extant species are at the 'top' of whatever branch begat them. We are no more 'advanced' than the humblest *Escherichia coli* bacterium, although, like the gorillas and chimps, we are certainly the brightest of the beasts. What we also are is self-aware (which is not necessarily the same as brightness; more of which later). Are we alone in this?

To get to the mountain gorillas of Rwanda's Virunga mountains entails some serious hiking. This is not hot, sweaty Africa, but surprisingly cool-and-misty Africa. Climbing up to the gorillas' lair is something like walking through the New Forest

– angled at 45°. It takes ages and you keep slipping and sliding through the mud, but it is worth it.

We were extremely lucky, that cold June day. We stumbled upon the Susa group, an extended family of some 30-odd gorillas, the largest single tribe of the animals (comprising, rather frighteningly, some 5% of the world's total extant population). There were a couple of large male silverbacks, several mature females, and some dangerous babies playing with the bamboo and celery.

The babies are not dangerous in themselves, of course. But we were told that they were to be avoided at all costs. Like all primate youngsters, they are mischievous, basically friendly and really only want to play. And if they do there can be trouble. 'A Japanese tourist made a mistake a few months ago', one of our guides told us. 'The baby came up, and he picked it up and held it. The silverback, ha! He didn't like this at all. He picked up the tourist – after taking back the baby – and threw him up into a tree. Broke a leg. Very nasty.'

In fact, it is something of a mystery why these animals are quite so powerful. They have, quite literally, the strength of 10 men, and are capable of snapping branches as thick as a leg. This strength seems to confer no obvious advantage on these animals. They are not especially aggressive towards each other, and have no natural predators except for us. It is either a hangover from a more red-in-tooth-and-claw evolutionary past (their dentition certainly says 'carnivore' rather than 'salad muncher', and in fact there is some evidence that gorillas are not pure herbivores to this day), or it is the result of some rather complex form of sex selection, a bit like the preposterous feathers of some tropical birds.

It is also a mystery why they are so bright (neither the environment nor the frankly catatonic lifestyle of the mountain gorilla is particularly intellectually demanding). But bright they most certainly are. After hanging around for 15 minutes or so with the group, they started to get bored with us. It was disconcerting to be this close to animals completely aware of and yet so utterly indifferent to human presence. Most species show either naked aggression or blind panic when *Homo sapiens* is around (quite sensibly, as it must have seeped into their brains that these nasty two-legged things will have their hide for a rug as soon as look at them). This mild curiosity plus studied aloofness is quite unusual. Anyway, a small group of the animals decided to wander off through the woods, and we decided to follow them.

The breakaway group consisted of, as I remember, two not-quite-mature females and a young male. They looked for all the world like a group of teenage friends going for a walk, and that is what, I suppose, they were. They padded almost silently through the bamboo, and led us to a small gorge, a tangled mass of spider webs and greenery under the canopy of the forest. In the gorge, a small but dramatic cleft in the mountain, was a stream, and in one place this widened out into a small pond, maybe three metres across. The three apes then sat down around the pool. One, a female I think, stared very intently at her reflection. I could swear she ran her digits through her hair, looking at her face in the water-mirror as she did so. Then one of the others, also staring at the reflections, jabbed a hand into the water, which of course broke up into ripples. At that point the three animals fell about, laughing at their now-wobbly reflections.

OK, these were animals. They were making sound, appropriate to their species, that consisted of a certain degree of whooping and whistling. A proper scientist, as opposed to a journalist or tourist, would no doubt describe their change in posture and body attitude using very different terms than 'falling about'. 'Who knows what is going through their minds?', scientists would say, so best not go there.

I'm sorry, but this will not do. Sometimes, if it looks like a duck, walks like a duck and quacks like a duck it is easier just to assume you are in fact dealing with a duck, rather than some sort of complex analogy. These gorillas were falling about laughing at what passes for entertainment in the Virunga forests. And if a sense of humour is not a sign of intelligence and self-awareness, it is hard to see what is.

Our attitude to animal self-awareness has historically been odd and self-contradictory, and at the heart of it are some very uncomfortable truths. The science of animal cognition has undergone something of a revolution in the past 30 years or so, and the findings are all pointing in one direction: the mental life of animals is far more complex and sophisticated than we thought.

Not only are animals cleverer than we once believed, they are probably also more emotional, more self-aware and in many ways more like us than we ever believed possible. Here, science is on a collision course with the world of accepted ethics and morality, and in the near future it is easy to see a revolution occurring thanks to what we are learning. If we decide that Fido is not a zombie, the entire relationship between humanity and the rest of the animal world will have to change.

Historically, as we shall see, the subject of animal rights has largely been a matter for theologians and philosophers. Latterly, it has been an issue for campaigners and activists. But today, the whole issue of what 'rights' we grant our fellow species has moved into the scientific domain. Not so long ago, anyone who suggested that other species could think, use language and tools, and show 'human' emotions such as love, kindness or empathy would have been accused of hopeless anthropomorphism and sentimentality.

Once, 'intelligence' in an animal was seen as purely 'instinctive', and 'instinct', however you chose to define that rather nebulous quantity, was one of those things that marked out a safe boundary between animals (whose every move, however complex, was deemed to be guided by it) and us (who, being 'higher' beings, are not so driven). The cleverness of species like dogs and chimpanzees has long been acknowledged, but until quite recently many scientists held these behaviours to be little more than party tricks, simulacra of conscious reasoning. They may look clever, the reasoning went, but this is an illusion. Behind those bright eyes is in fact nothing. Even the brightest animal is no more than a machine, a zombie.

But the mood has changed. The more zoologists study the behaviour of animals, the more complex and 'sentient' their behaviour becomes. Science has also come close to a quantifiable definition of sentience, a checklist against which we can measure the 'performance' of various species. And, inevitably, all this raises uncomfortable questions.

It has long been recognized that the great apes deserve special recognition for their intelligence and presumed sentience. Indeed, in many countries, such as Britain, species such as gorillas and chimpanzees have acquired a unique legal status, particularly regarding animal experimentation laws, that separates them from the rest of the non-human animal kingdom. But the more we find out about animal abilities the more awkward these questions become. Awarding 'rights' to chimps and gorillas is one thing, but what about dolphins? And if dolphins, what about other mammals, such as dogs and cats? Sheep and pigs? Crows? Fish? Hang on a minute: we eat some of this stuff. Unlocking the secret mental life of the beasts means opening a very nasty can of worms indeed according to some scientists and philosophers.

We need of course to first define what exactly we mean by *sentience*. There are seven 'markers' upon which scientists can perhaps agree. All are possessed by humans and some by many other species as well. A very few species seem to possess all of them.

This list of markers begins with a '*theory of mind*': the ability to know or guess what another being is thinking. A typical 'theory of mind' test would be to ask: 'What can that person over there see?'. Humans older than about four can do it; adult chimps and bonobos possibly. No other species demonstrate this highest-order cognitive skill (severely autistic people and young children of all abilities seem to lack a theory of mind).

Tool use, once thought to be the preserve of humanity, turns out to be very common. Various apes and birds, and even marine otters, can and do adapt natural materials to a variety of uses.

Plenty of species show evidence of strong *emotional and empathetic 'abilities'*, if that is the right word.

Another 'sentience' trait is the *ability to mimic*. In primates, neurons called 'mirror cells' seem to fire up when we try to copy others in performing tasks. Apes, obviously, can *ape*, as can (somewhat less famously) octopuses.

Language is certainly no longer considered to be an exclusively human trait and the '*mirror test*' – 'Do I recognize the being in the looking glass to be me?' – once seen as a key divide between the sentient and the 'zombies', has been passed by animals as diverse as pigeons and elephants (and it is questionable how good a test it can be for species whose visual abilities are far outstripped by other senses, such as smell).

Perhaps the 'highest' quality of sentience is *metacognition*, the ability to think about thinking. 'I think, therefore I am' was Descartes' famous summation of what it means to be self-aware, and until recently it has been our ability to ruminate, to live in a mental world apart from the world of the 'immediate now' that is assumed to constitute animal thinking, that has separated us from the beasts. That may be about to change.

Not everyone will be convinced. For, despite these advances, despite the papers in *Science* and other journals highlighting the case of genius crows and dolphins, despite the reports of extraordinary feats of sign language performed by some captive apes and even the apparently real linguistic abilities of some birds, there remains the whiff of pseudoscience about the whole field of animal cognition. It is, after all, impossible to really know what is going on in an animal's mind. We cannot, as the philosopher Thomas Nagel once pointed out, know what it is like to be a bat.

Sceptics – or cynics – like to say that research in this field is held back by an assumption that the plural of 'anecdote' is 'data'. Nevertheless, casting aside sentimentality, the undoubtedly dubious assertions of humanlike skills which have been made by some researchers, and the paucity of hard and fast information, there remains the important truth that the more we learn about animals the more like us, in certain respects, they are turning out to be.

This is new science. The evolutionary basis of intelligence and cognition in general is still very poorly understood. We do not know why humans became so bright. Our brains consume vast amounts of energy (about one in eight of every calorie we consume goes towards powering the computer in our skull) and their sheer size makes human birth a trauma not seen in most mammal species.

We tend to assume that our intelligence arose due to simple natural selection – the benefits of a sharp mind for survival seem obvious – but in fact the road to human intelligence may have been sparked by something far more 'trivial' – sex selection, perhaps. Our ability to gossip and form complex social relationships is mirrored by some other primates, but we have no idea why we, and not they, should have become quite so bright.

Of course, we *are* animals, but until recently it has been considered that intellectually *Homo sapiens* almost belongs in a separate kingdom. Now we cannot be so sure. And in a century when the roles and possible rights of animals, especially of species that are in mortal danger like the mountain gorillas, is likely to be thrown into ever sharper focus, then the science of animal sentience is likely to become more than a purely academic or philosophical debating matter.

Traditionally, the view of animal sentience was much influenced by religion, at least in the Middle East and Europe. Followers of the Abrahamic faiths held that the birds and the beasts are essentially chattels, ours to do with as we will. *Genesis* 1:26 states: 'And God said, Let us make man in our image, after our likeness: and let them have dominion over the fish of the sea, and over the fowl of the air, and over the cattle, and over all the earth, and over every creeping thing that creepeth upon the earth'.

This view has arguably dominated the whole issue of how most Westerners have thought about animals right up to the 20th century. In was never the only view though. In other, non-Abrahamic, societies animals can be viewed quite differently. In Buddhism for example every living creature is seen as part of a spectrum that includes human beings. Hindus see certain animals, particularly cattle, as sacred and will not eat or even harm them. But the Biblical view took hold in the society that ended up developing the sciences of evolutionary biology, ethology (the study of animal behaviour) and neuroscience. This rather uncompromising underlying belief about animals could be seen to have shaped our study and interpretation in a rather unhelpful way.

But even in Biblical times there were contradictions and paradoxes. Animals were treated badly, but religious codes arose forbidding cruelty. Animals were so much meat, yet in mediæval Europe they could be – and were – tried for murder.

Science, for once, took what can be seen as the traditional view. René Descartes famously asserted that all animals were automata, true zombies whose responses to things like pain were simply programmed reflexes.

He believed that only humans displayed sufficiently complex and refined behaviour to indicate the presence of a dualistic 'soul', a ghost in the machine necessary for consciousness. This idea was persuasive and it persuades still. An animal in pain, certainly a mammal or a bird which is in pain, appears to be suffering in the same way that a human who is in pain suffers. There will be screams, yelps, and the writhing of muscles, ligaments and skeleton suggestive of agony. We can monitor the animal's brain and its blood chemistry, detecting the presence of stress hormones like cortisol and adrenaline and note that its physical responses are identical to those of people.

Yet can we be 100% sure that the animal is actually *experiencing* pain in the same way as a human being? Of course not. It is quite possible to imagine a computer program or a robot designed to mimic the outward signs of pain, yet clearly there is no suffering to be had. I can quite easily program the machine on which this book is being written to shriek 'Ow!', or something like that, whenever I press, say, the letter 'Q'. Yet I'd have to be a moron to believe that my PC was actually in torment.

I'd also have to be a moron to assume that when an amoeba is challenged by nasty chemicals or intemperate heat or cold, and wriggles and squirms, that some meaningful perception of pain or even mild unpleasantness is going on. An amoeba is just a biological machine, a bag of proteins and nucleic acids, fats and water and various other bits and pieces no more likely to 'suffer' than my telephone or car. This used to be seen as a persuasive argument to 'prove' that animals are not conscious, but is not now generally accepted. An amoeba is as different from a dog as the latter is from a computer.

But not so long ago it was scientifically incorrect to argue that animals had a mental life at all. In the 1950s and 1960s, the behaviourists, a radical school of psychology, argued that just as it was a nonsense to argue that animals had a mental life, it was a nonsense, too, to argue that humans had one either. The arch behaviourist B. F. Skinner put pigeons in boxes and mapped out complex relationships between stimuli and responses, working on the premise that the birdbrain was a calculating engine.

Skinner trained his pigeons to perform extraordinarily complicated 'tricks', pressing sequences of levers in order to elicit a supply of tasty food and so on. What went on between the pigeon's ears was, he thought, not more than a continuation of these levers, an interconnected series of unknowing mental gearages that eventually made the animal do one thing or another. Skinner even tried the same trick with his daughter.

To the behaviourists, what went on inside the skull was unknowable and thus not worthy of study or even consideration. Discussion of the 'conscious' mind and what this might mean was like discussing fairies. Thoughts, such as they were, were at best merely an internalized form of language.

There is no doubt that behaviourism had a lot of useful things to say about how the mind works, and blew some useful mathematical rigour into the messy and colourful playroom of ideas that psychology was becoming. But there was a big problem. We all know we have internal, mental lives, because we experience them. Denying their existence because they cannot be meaningfully studied is like denying the existence of the Andromeda galaxy because no one has been there and probably never will. Of course, the behaviourist could take a solipsistic argument and assume that he or she was the *only* organism alive with a mental life, and his pigeons and fellow experimenters were mere zombies, but this would add unnecessary complexity to the argument (why should he be the only conscious being out of all the billions of others?). That fact alone would take a huge amount of explaining. Today you would be hard pressed to find someone taking a hard-line behaviourist view of animal or human consciousness.

But what does it mean to be self-aware and, indeed, conscious of anything? After all, it is perfectly possible, as a human being, to perform complex mental processes and not be conscious of them. If you drive to work along the same route every day, the chances are that during most journeys, most of the time, you will be no more conscious of your actions than you are of your heart beating or of your kidneys processing urine. Try to remember your trip the next time you end up in the office car park.

And yet, despite the fact that driving a car is a hugely complex and difficult mentally driven process that takes some time to master (and we can *all* remember what *that* was like), most of the time you do not crash in a heap of tangled metal despite having been a zombie for most of the journey. Many of our most complicated and impressive actions do not seem to be carried out under any sort of conscious control. We admire the 'skill' of elite footballers and tennis players, but what is it that we are really admiring when we see Roger Federer make an outlandish return of a 140 mph serve, or Ronaldo turning on a sixpence and blasting the ball into the net from 40 yards? After all, these actions must almost by definition be unconscious. The skill sinks in with the training, the hours of practice, and the guts and determination that is needed to be in the top 1% of any professional sport. Actually playing, at world-class level, is a spectator sport as much for the players as for the spectators.

If humans can be largely unconscious of their actions while playing tennis or driving, then chimpanzees can certainly be unconscious while hunting or grooming. But this does not mean that animals or humans are not self-aware. The question of self-awareness is one of the trickiest in science, but what is of the essence here is whether whatever-it-is is something possessed uniquely by *Homo sapiens*.

We cannot know what it is like to be a bat, a bird or a whale. But that is not to say we cannot study and discover some useful things about animal awareness. In 1970, the psychologist Gordon Gallup developed the 'mirror test' to determine whether animals were self-aware. In essence, the test uses a mirror to see whether an animal can recognize its own image as being one of itself. Plenty of animals are fascinated by mirrors, but to see whether they know the creature in the glass is themselves, the Gallup test involves marking the animal with a dye or paint mark (which it cannot see except in the mirror) and seeing whether the animal behaves in a way that indicates that it realizes that the reflected body, with the mark, is its own.

So far eight species have passed the mirror test, six unambiguously (humans, chimps, bonobos, orang-utans, dolphins and elephants) and a further two (gorillas and pigeons) under more controversial circumstances. Children under the age of two fail the test, as do (perhaps surprisingly) dogs and cats. One species of monkey, the capuchin, seems to be a 'borderline pass'.

The most recent alumnus of the mirror test was the African elephant. In November 2006, the *Proceedings of the National Academy of Sciences*[1] reported that three elephants, Happy, Maxine and Patty, who live at the Bronx Zoo in New York City, had spots painted on their foreheads and were then shown a mirror. All reacted in a way that indicated that they realized that the animals in the reflections were themselves. They poked their trunks into their mouths and watched the reflection in fascination. One, Happy, passed the spot test – she tried to wipe off the mark on her face with her trunk after spotting it in the mirror. 'The social complexity of the elephant', said Joshua Plotnik, one of the scientists behind the study, 'its well-known altruistic behaviour and of course its huge brain made the elephant a logical candidate species for testing in front of a mirror.'

The mirror test could be a good indication of self-awareness in species which are wired up in a certain way. But it is an exclusively visual 'test', and for many species vision is not their primary sense. Dogs, for instance, recognize each other mostly by smell. If some sort of smelly version of the mirror test were devised, we'd probably flunk. Neither we nor elephants nor chimps could pass a test based on echolocation, yet a bat might. The mirror test imposes a purely human (perhaps purely primate) criterion on the measurement of self-awareness. It still doesn't tell us much about what it is like to be a bat, a beagle or a badger.

What other criteria could we use to determine whether animals are conscious or zombies? There are a range of emotions that seem to depend on a sophisticated sense of one's place in the world and one's relation to it. Emotions like jealousy, sarcasm or humour seem to demand a sophisticated sense of self (less sophisticated emotions like love and hate, or fear and rage may not require anything like the same degree of mind). So, can animals get jealous? Can they be sarcastic? Most scientists are doubtful, but ask any dog owner and you will get the same answer. There are endless anecdotal stories of mutts creeping off to sleep under the bed in a huff when the new baby arrives, or naughty puppies 'hiding' socks and gloves behind the sofa and expressing great delight when their owners express suitable frustration. Much of this comes under the category of 'play' and, again anecdotally, there is a lot of evidence that many, many species engage in play. But how much of this is scientific and how much mere anthropomorphizing? And, most importantly, how much data do we have?

Not a lot. One recent anecdotal report claimed to confirm that jealousy can be experienced by dogs. The research, carried out at the University of Portsmouth in the UK, involved 1000 pet dogs and their owners across southern England. The owners reported many instances of jealousy, where dogs would become upset when affection was bestowed upon people or other animals.

The way in which this was expressed was usually by the dog forcing itself between its owner and the person it felt was usurping its emotions. Many dog owners report instances of their animals trying to place themselves physically between their master or mistress and new significant others, especially when they are being affectionate. Such behaviour is amusing, but may become less so when the significant other is a baby.

Persuasive though such stuff is, it isn't really science. Owners reporting on their dog's or cat's behaviour means little on its own; this is hardly a double-blind trial. Perhaps more persuasive are some neurological findings which suggests that animals may be able to fall in love.

In the brains of the great apes and in humans there is a structure composed of specialized neurons called spindle cells. They are found in the parts of the cerebral cortex which have been linked to social organization, empathy, sympathy, speech recognition, intuition about the feelings of others and emotional attachments. One area of these brain areas, the anterior cingulate cortex, seems to be associated with an emotional response to things like pain, sexual arousal and hunger. Another part of the brain, the frontoinsular cortex, generates a similar response when pain or suffering is seen in others. The strong emotional responses we get to other individuals – hate, fear, lust, love or affection – seem to depend in large part on the presence of these rapid-firing spindle cells.

And it seems that the great apes (including us of course) may not be unique. In what may be a classic case of parallel evolution, Patrick Hof and Estel Van Der Gucht of the Mount Sinai School of medicine in New York have found spindle cells in the brains of humpback, fin, killer and sperm whales, and what is more, they found a far greater concentration of them in the brains of cetaceans than in humans. Whales, Dr Hof told *New Scientist* in January 2007, 'communicate through huge song repertoires, recognize their own songs and make up new ones. They also form hunting coalitions to plan hunting strategies... and have evolved social networks similar to those of apes and humans'.

Perhaps the best way to determine sentience is the presence of abstract thinking. Humans can 'think about thinking', a skill called metacognition. It was this skill that was identified in Descartes's famous aphorism *'cogito ergo sum'*. Knowing what is, literally, on your mind would seem to be a key part of self-awareness. One traditional assumption has been that, lacking a language with which to internalize their thoughts, animals cannot do this. They can think about the pain they are in, but cannot worry about the pain to come. Lacking meta-cognition, animals could be said, if not to be zombies, then certainly to lack a key property of non-zombiehood, their inner worlds a clouded non-reflective series of stabbing consciousnesses.

But some researchers say that it is possible to unlock the inner lives of some species, and that they do show this property. A psychologist called David Smith, who works at the University of Buffalo in New York state, has been working for some years with a bottlenose dolphin called Natua in a harbour in Florida. He trained the animal to press buttons depending upon the frequency of the sounds it was hearing. When the differences between the sounds was obvious, the dolphin had no problem (a snack was the reward for getting the right answer). But as the sounds to be compared got closer in frequency, to the point where even the dolphin's impressive hearing apparatus is unable to distinguish between them, Natua learned to press a third button, effectively a 'don't know' or 'pass' button, that moved the test on to the next 'question'. Similar results were found with rhesus monkeys, this time using symbols in a computer game. The tests have been refined to determine the level of confidence that the animal feels that it has the 'right' answer. Smith told *New Scientist* in 2006: 'I can't claim these monkeys show fully-fledged consciousness, but I have shown the exact cognitive analogy to what we have in humans, and for us it is consciousness'.

Animals can be very bright. Apes, perhaps dolphins, and certainly some crows have astonished scientists and the public in recent years with displays of intelligence that were not anticipated by early researchers. Every year it seems we get new data that show that animals are probably cleverer than we thought. One of the attributes previously thought to be unique to humans was toolmaking. That notion went by the board as soon as it was discovered, in the 1980s, that chimps in east and central Africa could use moistened sticks to fish for termites. One West Lowland Gorilla in the Republic of Congo, a male named Kola, has learned to test the electric fence surrounding his forest reserve by holding a grass stem up to the wire. The stem will conduct a bit of current, enough to show Kola the fence is turned on, but not enough to give him a shock. Not all humans would be able to do this.

The fact that apes, our closest relatives, can be this bright is perhaps not surprising, but what has taken more than a few scientists aback is just how intelligent some birds, a group whose very name was previously a byword for stupidity, are turning out to be.

In the BBC television series *Life of Birds*, shown in 1998, some extraordinary footage was shown in which crows in Japan dropped hard-shelled nuts onto the road at a pedestrian crossing. After waiting first for the nut to be cracked open by a passing car and then for the traffic to be stopped when a pedestrian pushed the button, the crows would land to retrieve their nuts.

This is hardly very scientific. But In 2002, a New Caledonian crow called Betty starred in *Science*[2] magazine after she had learned to fashion a hook out of a piece of wire and use this tool to fish food out of a glass pipe. That really took scientists aback, especially as this level of toolmaking skill has never been observed even in chimpanzees. 'Primates are considered the most versatile and complex tool users', the authors of the study wrote, 'but observations of New Caledonian crows raise the possibility that these birds may rival nonhuman primates in tool-related cognitive capabilities.'

What was really extraordinary was that the hooks made by Betty were constructed from flexible steel wire, not a material readily available in the bird's natural habitat. Even more impressive was the way Betty made the hook. Crows lack hands, opposable fingers and thumbs. To make the hook, Betty first wedged one end of it in the sticky tape wrapped around the bottom of the glass tube and then pulled the other end at right angles with her beak. Betty had no prior training and had not watched any other crows doing this. Chimpanzees have, in similar experiments, shown themselves incapable of grasping the principle of bending a pliant piece of wire to make a hook and retrieve food. Some people would probably have trouble.

The fact that at least some birds are so clever comes as a surprise, partly because birds are so distantly related to humans. They are not even mammals. But of course there is no reason to suppose that 'IQ' (whatever that means in an animal context) should have any relationship to evolutionary closeness to *Homo sapiens*. In a review published in *Science*[3] in 2004, the question of corvid (crows, jays, rooks, magpies, ravens and jackdaws) intelligence was discussed. In fact, despite the pejorative term 'birdbrain', the smartness of these species has been known about for some time. In one of Aesop's fables, a crow, unable to drink from a pitcher of water because the surface was too low for its beak to reach, started dropping stones into the pail, displacing water until it was within reach. For a very long time, tales like this have always been dismissed as hearsay and folklore, which of course they are, but it is interesting that it has only been very recently that science has started to discover that many of the things lay people have thought about animal intelligence might be true. 'Recent experiments', the authors wrote, 'investigating the cognitive abilities of corvids have begun to reveal that this reputation has a factual basis.'

The authors speculate that intelligence evolves not to solve physical problems but to process and use social information, such as who is allied to whom and who is related to whom, and to use this information for personal gain and deception. This is all very well, but you need the equipment for the job, and here the crows and their relatives seem to tick this box as well. The crow has a significantly larger brain than would be predicted for its body size – in fact, on this measure it is similar to that of a chimpanzee.

Among the bird family, only some parrots have larger brains in relation to body size. The crow's brain is also particularly well developed in those areas thought to be responsible for 'higher' thought processes, a region of the brain dubbed the 'avian prefrontal cortex' as it is thought to be analogous to the structure seen in mammals.

Not only have crows been seen making hooks in captivity, their behaviour in the wild shows some extraordinary abilities. For instance, they cut *pandanus* leaves into a series of sawtooth spikes and use these to impale grubs and insects gathered from under vegetation. Many corvids store food for future consumption; they are not alone in this of course, but what is extraordinary is that they seem to be able to distinguish between perishable and longer-lasting supplies, and return to their cache before their stores have gone off. In laboratory studies, birds will not bother to return to a store of dead grubs, say, after a long time has elapsed, but will return to find seeds. This suggests a 'what, where, when' memory much like ours. These birds are also able to create complex strategies to cope with thieves. They hide their food where other birds cannot see it, or wait, their beaks stuffed, for other birds to fly away or turn aside before hiding their store. And birds who are habitual thieves tend to be better at hiding their food than ones who have not.

These birds, in other words, show flexibility, seem to be able to work out what others may be thinking, understand the principles of causality, show imagination and can plan. Most importantly, their behaviour strongly suggests that they are able to put this 'cognitive toolbox' together to construct an internalized image of the world. Does this mean they are fully sentient and conscious? We don't know. But it is good evidence, surely, that they are.

Betty and her relatives are impressive enough, but they cannot talk. Some birds are, of course, able to mimic human speech, but it is only quite recently that anyone has suggested that they are able to understand what they are saying. Alex, an African Grey parrot, has been studied for nearly 30 years by the animal psychologist Irene Pepperberg in the US. His vocabulary is about 100 English words, and he seems to understand what they mean. Pepperberg has claimed that Alex understands concepts such as shape, colour and material, and can use English correctly to describe these concepts. Alex also apparently shows remorse. He was even able to make up a word – 'bannery' – when shown an apple for the first time (he already knew about grapes and bananas). Another African Grey in the US, N'kisi, is, its owner insists (some scientists are skeptical about N'kisi), capable of using language to have a real conversation (a claim that has not been made about Alex). N'kisi is supposed to be capable of humour and even sarcasm. He invented a new term when faced with aromatherapy oils – 'pretty smell medicine' – although whether his sarcastic attitude extends to alternative therapies is not known. And it's not just parrots and crows. Sheep have now been found to recognize dozens of individual people. And earthworms have been spotted doing differential calculus (just kidding).

Cleverness is not self-awareness. The fact that crows can make tools out of wire does not necessarily mean that they are sentient. But intelligence may well be linked to consciousness. That the brains of mammals, reptiles, birds amphibians and even fish share common structures and genetic backgrounds suggests quite strongly that our self-awareness is almost certainly not unique. Because not to draw this conclusion would be to assume something very strange indeed, something along Cartesian lines – that somehow, at some point in the evolution of *Homo sapiens*, and *Homo sapiens* alone, something magical invaded our skulls in the Pleistocene and set up home.

So where does all this leave us? In an uncomfortable place, that's where. If we assume, as I think we must, that animals are sentient, aware beings capable of conscious thought and of distinguishing themselves from the rest of the world around them, if animals are not zombies, then the distinction between them and ourselves becomes somewhat arbitrary. As some people have argued, eating them is no different, morally, to cannibalism (indeed it is worse, because at least with cannibalism there is the possibility that the meal can give his or her consent to be eaten).

It is certainly uncomfortable for us that the more we learn about animals the more impressive their intellectual capabilities appear to be and the more, biologically, we seem to have in common. This can of course be overstated. It is an over-quoted factoid that humans and chimpanzees share more than 99% of our DNA, but it is a much less quoted factoid that we share two-thirds of our DNA with the halibut, and we are well into double figures with yeast. And anyway, what does this mean?

Humans and chimps may be on neighbouring twigs on the evolutionary tree, but mentally we might as well be in different forests. Nevertheless, things like tool-using, language, a sense (perhaps) of fairness, and even emotions like humour and jealousy – all once thought to be the preserve of people – have now been observed, to greater and lesser degrees, in animals. It is likely that the more we probe the minds of the apes and monkeys, elephants, dogs, birds and probably even fish, the more impressive will be the intellectual machinery we will discover. Even invertebrates are not immune to this animal 'intellectual revolution'; some cephalopods – the squids, cuttlefishes and octopuses – are so bright that they have won legal rights in some jurisdictions to be protected from certain painful experimental procedures.

It is incontrovertible that we equate sentience with both humanity and a right to humane treatment. Throughout history, the worst cruelties perpetrated by humans upon each other have often come about when the offenders persuade themselves that their victims are not really human and not really sentient. In the early 19th century, a famous advertisement was placed in a British newspaper for 'guns' (i.e. men with guns) to join a hunting expedition – to kill Tasmanian aborigines. In 1800 there were about 5000 of these people alive, but by 1867 they were all dead, reduced to a series of body parts displayed in museums. This shameful genocide was not considered as such by these 'guns' simply because the Tasmanian aborigines were not considered, by them, to be people.

Animals are not people, but then again if we are talking about intellectual ability (and sentience) many people aren't really people either. The critically senile, accident and illness victims in comas, the newborn – all have intellectual capabilities and sentience well below those of, say, an adult chimp or Kola the gorilla, yet across all societies and cultures the human will be granted far more rights under law than the ape.

As the Australian philosopher Peter Singer, who takes an extreme but persuasive reductionist approach to animal rights, argues, this is illogical. If it is right to take a chimp's life to save a human then it may also be right, under certain circumstances, to take a human's life to save a chimp. To argue otherwise is simply wrong and makes one guilty of arbitrary speciesism. The Great Ape project (the name of which, ironically, indicates a certain amount of arbitrary speciesism itself) is a loose group of scientists and philosophers which argues that we should extend certain legal rights to at least the 'higher' primates, our closest cousins, as a first step. This would mean that experimentation, for any purposes, even to test potentially life-saving medical procedures, should be banned on these animals, everywhere and in all circumstances. The great apes would in fact have similar legal rights in law as human beings.

If non-human primates can show evidence of *metacognition* – thinking about thinking and reflecting upon memories – that puts them in a wholly different light. They, and perhaps many other species, can no longer be thought to live in an eternal present, responding to hunger and pain, fear and pleasure with no concept of anticipation or reflection. It is, perhaps, one thing to cause pain to an animal which can neither anticipate nor reflect on its experience, but quite another to lead into a laboratory or abattoir a terrified creature which has already created a distressing mental picture of what is about to happen to it in its head.

Most people do not think like this of course. I know many scientists who argue, persuasively, that granting an ape 'rights' over, say, an elderly person suffering from Parkinson's

Disease is ludicrous. By experimenting on the brains of monkeys and apes scientists have made great strides in learning how this most debilitating and terrifying of illnesses works, and have taken steps towards finding a cure. Stopping this experimentation, in labs in Europe and America, has become the *raison d'être* of many animal rights organizations, some of which have resorted to terrorism to get their point across, and this we all agree is despicable.

And what about the 'rights with responsibilities' argument? Again, many scientists and lay people say, it is ludicrous to grant 'rights' to an animal that can have no concept of its responsibilities under the laws that are giving it protection. If we protect chimpanzees from vivisection, should we also not be prosecuting chimps when they murder each other, or indeed us (as happens)? Should gorillas get the vote? This is ludicrous, clearly, so perhaps we should think again about these 'rights' and dismiss the idea out of hand.

But actually the 'responsibilities' argument falls down very quickly. Because we grant a whole host of rights to human beings from whom we demand no responsibilities whatsoever. Again, we are talking about the very young, the very old, the sick, senile and the mad. Lunatics cannot vote and neither can babies, but we are quite rightly not allowed to stick electrodes into their brains in the advancement of medical science. Young children are absolved from full responsibility when they commit criminal acts. We have no problem in granting humans rights without responsibilities, so why not animals?

The best guess is that we will have to muddle through, perhaps tightening the cruelty laws a little, but essentially maintaining the same troubled relationship with the animal world that has held sway ever since we diverged from our closest relatives. But this state of affairs may not be able to continue forever. The more we learn about the most intellectually advanced of the animals, the more squeamish we will inevitably become. Every second humans kill some 16,000 animals for food – that is 50 billion lives taken per year. While this slaughter may be carried out in fairly humane conditions in wealthy countries with strict laws governing farm animal welfare, we can assume that the vast majority of these lives are ended in a relatively disgusting and brutal way.

It may well be the case that in decades or centuries to come we may look back upon the way we treat our fellow creatures today with the same sort of revulsion with which we now treat slavery – a practice which 250 years ago was widely accepted in most of the 'advanced' societies on the planet. This is not an argument for vegetarianism, but it is an argument for a lot more compassion.

So, where does the science go from here? While behaviourism, 'anti-mentalism', is probably defunct as a philosophy, behaviourist techniques have survived and the rigour of behaviourist thinking is, ironically, exposing the mental life of animals as never before. Scientists study – or try to study – animals both in the laboratory and, increasingly, in their natural environments as rigorously and methodically as if they were conducting a double-blind drug trial. This is not easy. When observing the behaviour of an animal as complex as the chimpanzee, for instance, over long periods, it is probably asking too much of even the most diligent human researcher to avoid drawing all sorts of emotional inferences about their subject matter.

The harsh truth is that highly intelligent animals are often extremely endearing and form close emotional bonds with their observers. But this doesn't mean that the extraordinary fieldwork such as that conducted by Jane Goodall with 'her' chimpanzees has not added hugely to our knowledge of these extraordinary animals. More and more ethologists want to study the cognitive abilities of apes and cetaceans, animals too large, demanding and expensive to observe in any numbers until quite recently. Controversially, some scientists have tried to push animal cognitive abilities to the limit – trying to teach chimps sign language, for example.

Clever new techniques have been used to unlock the animal mind. The Gallup mirror test, while imperfect and probably not definitive, is giving us startling new insights into animal consciousness. It is now considered that what may be a defining characteristic of true sentience – a theory of mind, or 'knowing what the other guy is thinking' – is possessed by at least some primates. Finally, there has been the growing cultural awareness of what science has known since the 19th century: that humans are animals. The biochemical, neurological and evolutionary relationships that led to our minds and the minds of other species are now being mapped. Hardly anything now is considered to be definitely, absolutely, 'unique' to humans. Intelligence, tool use, language, fear, jealousy and anger have all been observed in many species.

Chimpanzees have been observed engaging in behaviour that it is hard to interpret as anything but extremely violent, vindictive and even sadistic. Even the humble rat has been found to be alarmingly 'humanlike' in many of its traits, displaying signs of affection, bloody-mindedness and even addiction to various narcotics (it is quite easy to make many animals alcoholics, nicotine addicts or even persistent and enthusiastic users of cocaine). While many animals may mirror 'our' finer cognitive traits, they are not immune to our baser ones either. The old line about it being unfair to compare thuggish people to animals because 'animals never stoop that low' is not correct. In essence, the study of animals' brains has become more like the study of human brains and vice versa.

This remains a hugely controversial field. Behaviourist thinking survives, and acts as a useful antidote to those who see evidence of profundity every time a dog barks or a whale flips its tail. The attempts to teach animals to 'talk', or at least sign, look, say sceptics, far more impressive in TV documentaries than in the cold hard light of laboratory trials. Perhaps inevitably, this is a field which attracts flaky thinking like moths to a candle – to some, it is only a short step from talking parrots to telepathic parrots.

We may no longer be alone, and this will, inevitably, affect the way we treat our fellow consciousnesses. Hurting a zombie is fine because the zombie cannot mind. But how many scientists now believe that even their rats are zombies? For the moment the mainstream scientific establishment considers that it is, just about, OK in extreme circumstances to experiment on a chimpanzee. Will this still be the case if the chimpanzee asks us to stop?

References

1 Plotnik, J., de Waal, F. B. M. and Reiss, D. (2006) Self-recognition in an Asian Elephant. *Proceedings of the National Academy of Sciences*, **103**(45), 17053–7.

2 Weir, A. A. S., Chappell, J. and Kacelnik, A. (2002) Shaping of hooks in New Caledonian crows. *Science*, **297**, 981.

3 Emery, N. J. and Clayton, N. S. (2004) The mentality of crows: convergent evolution of intelligence in corvids and apes. *Science*, **306**, 1903–7.

2

why is time so weird?

Time makes our lives. It is the key to how we perceive everything, from the ticking of our own minds to the events which mark our passage from birth to death. We can perhaps imagine a universe without colour, or without heat or light, but we cannot imagine a world without time. And yet, as far as physics seems to understand it, we may have to.

When it was assumed that base metals could be turned into gold, it was naturally assumed that there must be a substance which could effect this. Now, like the old philosopher's stone, we assume that there is a 'quantity' which marks the passage of events. Just as space stops everything happening in the same place, time stops it all happening at once. But while we know space is there – look, I can wave my hands through it right now – time is qualitatively different. We cannot, after all, wave our hands through time.

The true nature of time continues to elude us. Physicists have made huge strides in the last century or so in the way we think about time, but as to what it *is* exactly we are not really any wiser than the Ancient Greeks. Plato, after all, thought time was an illusion, and his view seems to be coming back into fashion. As the mathematical physicist Paul Davies says, 'Nothing in known physics corresponds to the passage of time... how can something so basic to our experience of the physical world turn out to be a case of mistaken identity?'.

We talk of time 'flowing', but flowing through what? At what speed does it flow and why? And what is the 'substance' that flows? A spacecraft moves through space and its motion can be described relative to other objects. But the passage of time cannot be described with reference to anything other than itself.

We can think of time in terms of mathematics and physics, and we can also think of it in terms of perception. The way we think about time seems to be linked in a rather odd way to the way we think. We do not really 'see' the passage of time, after all. We simply have a series of subjective experiences that are different from the ones in our memories and it is this difference that our conscious brains perceive as time.

Time throws up all sorts of paradoxes. You can use the existence of time, for a start, to prove that nothing is real. The past is as dead as those who no longer live, no more real than your dreams, right? And the future has not happened yet. So all that is to come is, again, imagination. All that is real, therefore, is that infinitesimal sliver of time between past and present, which of course amounts to nothing, because as time never stops that sliver has zero thickness. So, time is real, but nothing else is.

Since Einstein, many physicists have tried to break away from the traditional, common-sense view, or *presentism*, i.e. the view that it is only the present, the world at the moment 'now', that exists. New ideas about time recast this slippery quantity and try to remove the subjective perception of time by human brains from the equation, as the presentist view of time is increasingly at odds with the way physicists believe the world to be.

?

What science has always been happy to do with time is to ignore the philosophical horrors it throws up and just get on with measuring it, giving it a symbol and plugging it into our equations, represented by a nice little letter, like t, doing its job, oiling the clockwork of the spheres. Time is a fundamental quantity, meaning that it cannot be defined by reference to any other quantity. We can only measure it and use time to derive less fundamental quantities. A change in velocity over time gives us acceleration. If you drop a pebble down the well, you can calculate the depth in metres simply by multiplying the square of the time, in seconds, before you hear the splash by five (if this is Earth you are talking about).

Einstein showed us that the pull of gravity and the tug of acceleration were equivalent (the force you feel in an accelerating rocket is exactly the same as the force you feel as your weight is pulled towards the Earth). Indeed, Einstein went on to show that 'space' and 'time' are really different sides of the same coin. Before Einstein, it was thought that space was filled with an invisible medium called the ether, waves in which carried light and other electromagnetic radiation just as air carries sound. All that was needed was to work out the properties of the ether, particularly how it deformed and responded to the input of energy, and physics would be solved.

But the idea of ether had to be abandoned when it was found (in 1887 by Albert Michelson and Edward Morley) that the speed of light, as measured by an observer, is the same regardless of of the speed of that observer relative to the source of the lightbeam. Later, the Irish physicist George FitzGerald and the Dutchman Hendrik Lorentz suggested that this could be explained if you assumed that time actually ran more slowly in moving objects, meaning that light would appear to travel at the same speed for everyone. But they still assumed that this movement was relative to an ether.

But in Einstein's relativity, the old ether was abolished and replaced by space–time, a sort of conceptual super-ether, through which motion and the attracting force of gravity can be plotted. Because it was impossible to measure one's velocity relative to the ether (the old ether) then the whole notion was flawed. Instead, we all have our own personal time (in fact each independently moving point in space–time has its own personal time; my left foot's is slightly different from that of my head). If you take Einstein's ideas about time to their logical conclusion (as indeed he did) you must throw away all ideas about 'rivers' and indeed of pasts, presents and futures. Plato would have approved.

One can plot an object's movements through space–time as one can through the three spatial dimensions. For example, you can represent the orbiting of the Moon around the Earth as a helix, with time forming the vertical axis. Physics sees time very much as a 'label', a way of thinking about events, and especially the relationship with events that can be described mathematically. A point, for instance, occurs in both space and time. If you take two points, P1 and P2, they can have a series of different relationships, spatially and temporally, to each other. But their spatial relationship is qualitatively very different from their temporal one.

If P1 occurs before P2, for instance, then P1 can affect P2. But P2 does not have the same relationship with P1. You can talk about the 'causal future' of P1 in other words in the same way that you are not able to talk about the 'causal past' of P2 – simply because (in our Universe) events do not have 'causal pasts'. Einsteinian space–time in fact delineates the geometry of both space and the order of events. From any occurrence, the effects of that occurrence throughout subsequent time form a 'cone', extending into the fourth, temporal, dimension. Within the cone, all is causal and logical. Without, causality is lost and madness reigns. This ordering, the idea that time is a way of saying that one thing follows another as a *result* of the other, seems to be the key to time's true nature.

Einstein made us throw away any idea of absolute simultaneity. Two events which occur at the 'same time' in one reference frame may occur at different times when viewed from another. The question 'What is happening on the Moon right now?' actually has no meaningful answer. Someone on Earth is always a second and a half away from being able to know anything about what is happening on the Moon, as no information can travel faster than light. As there is no absolute, privileged 'now', it's better instead to think of a 'timescape' where time is laid out in its entirety.

The way in which time works seems to run counter to the way just about everything else in physics works. For instance, impulses can be reversed: objects slow down and speed up. For most things, there is no arrow, no one-way street. Processes are symmetrical. But time-dependent processes are different. If you drop a glass on the floor and it breaks, the pieces do not – and will never – make themselves back into a new glass again. Entropy, the degree of disorder in a system, will tend to increase over time (the second law of thermodynamics). The 'one-wayness' of time is proving a real headache for physicists. It is quite probable that we will need to discover a lot more about time before quantum physics and relativity can be reconciled into a theory of everything.

Unlike quantum effects, time is something we perceive directly. We have memories of the past but not of the future. Neither the future nor the past are 'real' in the sense that they are accessible and measurable, but one seems to have a privileged position over the other: the fact the past has 'happened' gives it a reality denied the future. The Big Bang seems to have imposed a cosmological arrow of time. The galaxies (or rather superclusters of galaxies) have been flying apart for around 13.7 billion years and there is now no reason to suppose this situation will ever reverse itself.

Perhaps the opposite extreme from thinking of time as nonexistent is to treat it as a fundamental – perhaps *the* fundamental – component of the Universe. A theory called loop quantum gravity, one of the attempts made to resolve the differences between quantum theory and relativity, suggests that fundamental particles may be composed of tiny rolled-up braids of space–time. It is an attractive concept – at the heart of it all is just space and time, no 'stuff' at all. In the words of science writer Davide Castelvecchi, 'if electrons and quarks – and thus atoms and people – are a consequence of the way space–time tangles up upon itself, we could be nothing more than a bundle of stubborn dreadlocks in space'.

The trouble is that, pretty though this idea may be, like string theory it is very hard to test experimentally. To detect strings we may need to build particle accelerators the size of a continent in order to generate the necessary energies.

To probe the internal structure of these strings we would probably need to start building planet-sized atom-smashers in space. Since it is a fair bet that NASA's or CERN's budget will not run to this any time soon, we have to rely on theory coming up with the goods. That or perhaps – just perhaps – getting lucky and detecting what some physicists think may be remnants of the colossal energies of the Big Bang: cosmic superstrings, giant expanded subatomic particles billions of miles long made of 'stuff' so dense that a metre of superstring would weigh trillions of tonnes.

?

Time as a fundamental quantity seems to be intrinsically linked to our conscious perception of the world. The philosopher Derek Parfit, in his seminal 1986 book on personal identity, *Reasons and Persons*, wrote of 'the objectivity of temporal becoming'. Time is, he says, linked intimately and extremely weirdly to our notion of continuous personal identity, something to which most of us give little thought, although it is perhaps the key aspect of our existence.

The idea that time is just the fourth dimension of space, one which we have a special interaction with through the offices of our conscious minds, is an attractive one. And clearly there is an element of truth in it. It is impossible, as H. G. Wells's Time Traveller said, to have an 'instantaneous cube'. One cannot exist for an infinitesimal amount of time.

The nature of time does seem to be very tied up with the way we perceive it. The Universe that we perceive is, in the most fundamental way, the creation of our sensory apparatus. When you touch a wooden table top with your fingertips you are actually 'feeling' the tiny repulsive forces of trillions and trillions of electrons; you are not actually 'touching' anything at all. When you 'see' objects you are simply making sense of reflected and emitted photons which your brain has, in a probably rather arbitrary and contingent way, evolved to interpret as light and dark, colours and so on.

It is the same with time. The strange and elastic nature of perceived time is well illustrated by a quote from Albert Einstein: 'When a man sits with a pretty girl for an hour, it seems like a minute. But let him sit on a hot stove for a minute and it's longer than any hour. That's relativity'.

This quote is lifted from the abstract for a short article Einstein wrote for the *Journal of Exothermic Science and Technology* in 1938 (he went on to detail the difficulties of obtaining a hot stove and a pretty girl – 'I live in New Jersey'). Many people have remarked that time can sometimes appear to flow at dramatically different rates from 'normal'. For example, it is common to report that time dramatically 'slows down' just before an impending disaster, such as a car collision or being thrown off a horse.

Professor David Eagleman, a neurobiologist at the University of Texas Medical School, devotes his research to the perception of time. He illustrates a very fundamental anomaly with the 'flash-bang' phenomenon. Sprint races are started by a bang rather than by a flash because our brains (or at least the unconscious motor-response bits of our brains that sprinters rely on to get them off the blocks) respond to sound faster than they do to light (even though sound travels to our ears at only a millionth of the velocity of light waves).

When it comes to perception the conscious parts of our brains perform a very clever editing trick to make sure that we think sound and light travel at the same speed. If you click your fingers in front of your head, you perceive three things as happening simultaneously: the noise of the click, the sight of your fingers clicking and, most weirdly, the actual decision to click them right at that point. In fact, these events all happened at different times.

You 'decided' to click your fingers several tens of milliseconds before the nerve impulses were sent down your arm to enable the muscles to do so. The sound of the click arrived at your brain an instant behind the light. These are tiny amounts of time, but our brains are capable, if that is what we want them to do, of perceiving extremely small time intervals, of the order of hundredths or even thousandths of a second.

It is quite possible to trick the brain's temporal editing suite into mixing up time altogether. In an experiment published in 2006, Professor Eagleman set up an experiment in which volunteers were told to press a button[1]. When the button was pressed the light would come on almost instantaneously as a result. At other times, they were asked to press the button after the light came on. In all cases the subjects were able to distinguish perfectly the order of events.

Then, the experiment changed. There was now a 100 millisecond delay (an easily perceived time interval) between pressing the button and the light coming on. But after pressing the button a few dozen times, the brains of the volunteers recalibrated the time delay back down to almost zero. But now comes the interesting bit. The time delay was changed from 100 milliseconds to 50 milliseconds. What happened now was unexpected. In some cases, the subjects perceived that the light came on *before* they pressed the button. In other words, their brains had been tricked not only into misperceiving the order of events, but misperceiving causality as well.

One of the most unsettling experimental results in the history of science was the study by neuroscientist Benjamin Libet. He monitored brain and nerve impulses to show that the movement of muscles under supposedly conscious control – the tapping of a finger, for example – is governed by nervous impulses from the brain and spinal cord *before* we become aware of wanting to move. This is disquieting, as it strongly implies that our sense that we are in conscious control of our actions is an illusion.

In fact, our perception of time shows us something quite interesting about consciousness and free will. We neither decide to act then act, nor predict that we will decide to act then act, but instead construct a *post hoc* fiction that 'we', meaning our conscious minds, made a decision.

And time perception is a key part of all this. Eagleman says: 'I often wonder what, if we could work out what was really going on in reality, and we were able to take time out of the equation – that little *t* embedded in so many physics equations – would this have an emotional effect?'.

So much of our emotional lives is tied up with a very singular perception of time. Grief, loss, apprehension and anticipation are all tied up with an internal model of time and causality. Einstein wrote to the sister of his deceased friend Michel Besso: 'That Michel has preceded me from this strange world is not important. For us physicists the distinction between past, present and future is an illusion'.

When someone dies we think of them as irretrievably lost, yet they existed for long enough to have made an impression on our consciousness – enough for us to miss them. You do not grieve for children before they are born, yet the Universe a week before the birth of an infant is as bereft of that person as the one a week after they die, 80 years hence. Time in this way has such a close relationship to our experience of it that this, perhaps, sets it apart from all other physical quantities.

In fact, it is quite easy to imagine a way of living that perceives time in a very different way from the way in which we humans see it. We cannot see into the future – that seems to be forbidden – but we can certainly see into the past. We live in the present – that is the way we have evolved and it clearly works best for us. But imagine if you were a creature for whom the past was equally 'real' to what was happening now. Events that happened a day, a month or a decade ago would be just as real as events that are happening now. Of course we do operate a bit like this – we live constantly not in the true now but in the now of a few seconds or milliseconds ago. It takes time for us to catch up with time. But for us the past is dead.

If it were not then we would have a very different view of the Universe. Death, for instance, would lose its impact somewhat. If the lives of those we had lost were as real as the lives of those still with us, would we mourn the dead like we do?

Derek Parfit wrote that it would perhaps be better for us if we were 'timeless', treating both past and future with equal weight. Then, we would not, it is true, be relieved when a bad thing is over, but neither would we be sad when a good thing is past. We would, in a real sense, be cheating time. For a timeless person, ten hours of agony yesterday is as bad as ten hours tomorrow. But for him, death also holds no more fears than birth.

Curiously, we seem to be better at taking a timeless view of others; suppose we are told that a close friend who lives abroad is very ill and is dying. We hear they face three months of misery before the end is predicted to come. Then we hear that there was a mix-up. The friend *was* ill, but that was some time ago. There was indeed three months of suffering but the friend is now dead. Do we feel better about this? No – which is perhaps strange, as we would feel better if we were told our own suffering was over and not yet to come. Because it is our friend we are unhappy at the idea of their distress, whenever it happens. We are able to treat our friend's past, present and future with rather more objectivity than we do our own lives.

'In our everyday experience', writes Piet Hut[2], a physicist at the Institute for Advanced Study at Princeton, 'time flows, as we flow with it. In classical physics time is frozen as part of a frozen space–time picture. And yet there is as yet no agreed-upon interpretation of time in quantum mechanics. What if a future scientific understanding of time were to show all previous pictures to be wrong and demonstrate that the past, the future and even the present do not exist?'.

Causality, the key to time, seems to be something hard-wired into our brains. As Toby Wiseman, a physicist at Imperial College London puts it, 'maybe deep within our brains we have this notion of cause and effect. Maybe no physicist could come up with a theory where causality didn't work that way'.

The idea of 'block time', an Einsteinian or Platonic time landscape where past, present and future are all equally real, is popular amongst physicists as it discards the apparent subjectivity of the passage of time. Nevertheless, the idea that the future has, in a sense, already happened and that we should no more fear our death than our birth runs so counter to our commonsense view of the way things are that it will probably never be accepted as folk science. It is an intriguing thought that if we *could* break away from our presentist view, then maybe we would be a lot happier. We would lose both fear and regret. We would also lose anticipation. So there would be a price to pay for being timeless.

It is certainly hard to imagine a universe where time worked in a radically different way from the one we are used to. All sorts of alternative universes have been postulated – ones where the gravitational constant is different, or where the weak and strong nuclear forces have different values from those in our Universe. Most of these universes would be very different places and ones very challenging to life. It seems to be the case that if you change the physical parameters even by a bit you end up with a universe that is either a massive black hole or a boring sea of elementary particles.

At least these alternative universes are respectable and polite entities, albeit rather dull and unexciting ones. But there is one constant that you cannot change, and that is causality. If you muck around with the order of events you end up with a very badly behaved universe indeed. Space, mass, the forces and the menagerie of particles are all essential parts of the recipe of being. But time, it seems, even if we are not very sure what it is, really is of the essence.

References

1 Stetson, C., Cui, X., Montague, P. R. and Eagleman, D. M. (2006) Motor-sensory recalibration leads to an illusory reversal of action and sensation. *Neuron*, **51**(5), 651–9.

2 Hut, P. (2006) A radical re-evaluation of the character of time. In *What Is Your Dangerous Idea?* (ed. J. Brockman). Simon & Schuster, New York.

3

can i live forever please?

Ageing is as inevitable as the sunrise and taxes, right? Well, yes, if you are unlucky enough to be born as a human. We get made, we live our lives, we wear out, we croak. It's sad and depressing and there is not much we can do about it, so we might as well get used to it. 'I don't want to achieve immortality through my work', Woody Allen once said. 'I want to achieve it through not dying'.

Someone who thought very much along the same lines was the late Professor Roy Walford, a biologist at the University of California Los Angeles. Walford believed he had discovered the secret, if not of immortality, then certainly of the key to a much extended lifespan. According to Walford, the secret to life extension was to eat less. Actually, to eat almost nothing.

I visited him once, back in 1999, at his house in Venice Beach. At the time he was in his early 70s and had the general demeanour of an elderly Hell's Angel rather than of a respected academic: elaborate facial hair, bald pate, denim shirt, medallion (I think) and an unexplained pair of attractive women scuttling through the house. We sat down and he explained his philosophy.

'I think I can get by on less than a thousand calories a day', he said. Walford was the world's leading proponent of calorie restriction as a way of extending life. He wrote some popular books (*The 120 Year Diet*, *Beyond the 120 Year Diet: How to Double Your Vital Years*) and set up a web site, and a sort of online club came into being, a low-cal community that followed Walford's dietary advice. 'Please forgive me if I eat', he said, and his 'supper' was brought, a bowl of rice and some water. 'Don't you lose the will to live, eating like this?', I asked him. 'Nope. The more I eat like this the more I'll live and that's that', he replied. After I finished the interview I went off and searched out a pizza.

A couple of days later I ran into Greg Stock, a UCLA biologist colleague of Walford's and another scientist who thinks it may be possible to do something about ageing. As we sat down to tea and cakes in the UCLA refectory (Stock is no calorie restrictor), he told me about his plan to organize a big cash prize for anyone who came up with a concrete way to extend human lifespan. The prize remains unclaimed. Just before we parted he asked me how Walford seemed. 'He's amazing', Stock quipped. 'Must be 72, 73... and doesn't look a day over 80!' Walford died in 2006. Woody Allen is still very much with us, but sadly we have no reason to assume his wish will be granted any more than it was for Walford.

Ageing is a very odd subject. Like the nature of time it is something we still have not entirely come to grips with. Like the nature, and disputed existence, of animal sentience it is political and hugely controversial. There is a lot of work being done about the nature of the ageing process, but rather less on doing anything about it. Combatting ageing is one of those areas where there is a huge, yawning chasm between public perception and desire, and what scientists are actually doing and prepared to contemplate. Anyone who suggests they have found a way to extend human lifespan is usually dismissed out of hand as a nutter. This is not surprising, as to date date the field of life extension has been liberally peppered with the eccentric and insane. Perhaps more surprising is the opprobrium heaped upon anyone who even *wishes* it were possible to make us live longer. There are a few scientists of ageing – biogerontologists – out there who regard ageing as a 'disease' and therefore, at least in principle, curable. But these people are not popular in the scientific community. There is a paradox here. Death is not popular. Yet we don't seem to have much of a will to do anything about it, even in the scientific community, which possibly could. This is odd.

How long you are likely to live varies hugely among species. It might have been better for both Walford and Allen to have been born, say, tortoises, which have been reliably recorded to reach nearly 200 years. Or better still, a whale. Whales are brighter than tortoises (although we don't know if they have a sense of humour). Mariners used to tell tall tales of the great cetaceans reaching incredible ages – 150, 200 years and more. These stories were until recently dismissed as tall beyond belief. But bowhead whales have been found recently with antique harpoons embedded in their skulls – weapons dating from the 1790s. It is therefore possible that there are large sentient animals alive today that are older than the US. Georgian whales: extraordinary. Some deep-ocean fish have been found that may live well into their second century. The Orange Roughy, a species now popular as food (it was previously known by the less-appetising name of 'Slimehead') does not mature until its mid-30s, and the individual on your plate may have been swimming in the Pacific when Lincoln was in the White House.

There are some suggestions that some of the very large dinosaurs may have been, if not immortal, then true Methuselahs, even compared to today's longest-lived whales, tortoises and fish. The great sauropods, some of which may have been 200 feet (65 metres) long, were probably immune (once they reached adulthood) to any predators, and infectious diseases and cancers may have progressed very slowly. It is not wholly impossible that some of these great animals lived a thousand years or more. Some sea anemones and molluscs appear to be effectively immortal, in the sense that they do not weaken and succumb to diseases of ageing, but how long actual individual animals can live in the wild is moot. To do better than this you need to leap out of the animal kingdom.

The bristlecone pine is one of the regular attractions in *Guinness World Records*. One specimen, alive and well-ish in California, is 4844 years old (as determined by counting the rings in its trunk). Several trees have maximum lifespans of a thousand years or more, including the giant sequoias. This is hugely impressive, but the true record holder is several orders of magnitude more impressive again. In 1999, some bacterial spores were successfully revived in a laboratory. These specks of life had been found encased in salt crystals in a cave in Carlsbad, New Mexico. Nothing strange about that – except that the spores were estimated to date from a quarter of a billion years ago. That means there are bacteria alive today that predate the first dinosaurs.

At the other end of the scale life can indeed be short. Some insects (mayflies are the most famous examples) have adult stages that last but a day or two (although their total life cycle is rather longer). Small mammals – mice, shrews and so on – are lucky to make two years. Birds tend to do rather better, as do bats. Humans come in near the top end of the scale, with the oldest recorded person to date, the magnificent Jeanne Calment, reaching a scarcely credible 122 years and 164 days. Mme Calment was French, and thus took little heed of a 'healthy lifestyle'. Indeed she smoked well into her 90s and ate and drank pretty much what she wanted. It is possible that there have been older people, but not many. Tales of remote mountain folk living to fantastical ages are just myths – the one thing all these places have in common is that the registrar for births, marriages and deaths is an infrequent visitor.

So what does all this tell us? Do, for instance, ageing and death come to us all? Those whales and tortoises die eventually, sure, but what about those bacteria? Two hundred and fifty million years is an extraordinary amount of time for an object as complex as a living cell (complete with genetic material) to be preserved (there is some controversy as to whether the revived bacteria were still capable of respiration and division). The existence of these objects shows that true immortality is probably not unattainable.

There is a great deal of variation in lifespan, and it is not always clear that there is a pattern. Generally big things tend to live longer than small things – this goes for plants as well as animals. Mice and the like live a couple of years; insects far less; dogs, rabbits and cats rather more. Fish, unexpectedly perhaps, live quite a long time. Even the little goldfish can live to nearly 50. Large mammals and reptiles do well – elephants can live into their 70s and humans, as we have seen, well into their 100s, although our great ape relatives are lucky to reach three score. Horses and cattle, despite their size, do badly. And bats and birds, despite theirs, do very well.

And slow things tend to live longer than fast ones. Those long-lived deep-ocean fish lead sluggish, almost catatonic, lives compared to their frisky near-surface cousins. If you hold a shrew in your hand its heart beats so fast it almost vibrates. Big animals, like whales and elephants, have a much slower metabolism. There is some truth to the old idea that while lifespans of different species may vary wildly, the number of heartbeats in an animal's life tends to be about the same.

When it comes to human ageing there are more patterns. Generally we have been living longer and longer as time has passed, but it is not clear that this has always been the case. A strange misunderstanding has grown up, by the way, concerning the term 'average life expectancy' and what this really means. For instance, in Ancient Rome the average life span of a citizen was less than 25. This has led many people to suppose that people in their 20s and 30s were 'elderly'. I have heard TV commentators asserting that in African countries today, where life expectancies may be in the 40s or less, people in their 30s are 'old'.

Of course people in Ancient Rome did not consider themselves 'old' in their 20s, no more than Africans aged 35 do today. An old man then, as now, was someone with grey hair and a grey beard with at least five decades behind him. These frighteningly low life expectancies had very little to do with ageing and everything to do with the ghastly levels of infant mortality in most pre-modern societies. If three babies out of five die before their first birthday, the survivors could all reach 100 and 'average life expectancy' will still be only 40. If you strip infant mortality out of the equation (by working out life expectancy aged five, say) children in classical times could probably expect to live into their 50s or 60s, and many into their 80s and beyond.

Most of the advances in life expectancy in recent centuries have been down to reducing infant mortality rates. Clean water, an understanding of sanitation and diet, and in particular inoculation and vaccination have, in the past 150 years or so, effectively doubled the amount of time a wealthy First World human can expect to live on emergence from the womb. A few decades ago it was assumed that life expectancy in developed countries would reach some sort of plateau around now, but that has not happened. In fact, in the world's wealthiest nations, average life expectancy continues to increase at around two years per decade – which means that in the richest countries, every day pushes up life expectancy by five hours.

What is rather less well known is that until the advent of all these wonderful new technologies, human life expectancy had probably been *dropping* steadily over the past thousand years despite – or rather because of – the vast advances in technology and in particular agriculture made by *Homo sapiens* since the last glaciation. It is now believed that our Stone Age ancestors may have enjoyed rather longer lives, even taking into account infant mortality, than, for example, mediæval Europeans. (They were rather taller, too.)

There is some evidence that in today's poorest countries those people maintaining a traditional hunter–gatherer lifestyle tend to do rather better than those at the bottom of the agricultural heap, and there is good reason to suppose that with the advent of agriculture and the specialization of industry the general health of the population went down, as the price we paid for an increase in the number of people that a given parcel of land was able to sustain.

Our diets became blander and less nutritious, we introduced more refinement to our food (milled grains) and, in particular, by living in larger, fixed communities, we became much more susceptible to epidemics of infectious diseases. It wasn't until the advances in sanitation and understanding of how diseases spread that the 'wealth' allowed by the Great Leap Forward into agriculture some 10,000 years ago paid the dividend of a healthier, as well as a larger, population.

In some other ways we are still going backwards. In the 'Stone Age', a loose term which I am using here to indicate any pre-agricultural civilization, humans may not have been very long-lived and may well have been constantly on the verge of starvation, but they were superbly fit. Most of today's remaining hunter–gatherers are quite happy covering 30 to 50 kilometres a day, running or walking on foot. Hunting, and even the gathering of shellfish and berries, was and is an intensely physical activity. The average Palaeolithic European or Neanderthal probably expended twice the calories per day in exercise as the average Roman citizen, and the average Roman three or four times as much again as we do today. It is interesting to speculate what our lives, health and life expectancy would be like if we could combine a Palaeolithic hunter–gatherer lifestyle with all the advances of modern medicine and sanitation. One suspects that we might well see a substantial jump in life expectancy.

It is quite easy to see how we could increase life expectancy, but it is not at all easy to see how we could increase our maximum lifespan. This seems to be set at something around 120 years. Even the healthiest populations, shielded from epidemics and violence, made up of people with the resources to eat well and take advantage of medical care, seem to be unable to produce more than a handful of 'supercentenarians': those who live to 110 or more. To date, the world has seen some 800 documented examples.

Compare to this the near one million ordinary centenarians in the record books. While it is certainly becoming much easier to live to be a hundred (in the 1960s there were on average only about 300 centenarians alive in the UK; forty years later the figure is around 6,000 and by the late 2030s the prediction is that this will increase tenfold again – in a population that is relatively stable), it does not seem to be any easier to live to 112. In 1857, there was one 112-year-old alive in the world, a Dutchman called Thomas Peters, who died that year. Today, there are about a dozen or so people alive at that age or above, but between the 1850s and now the figure has at times been higher. Since Jeanne Calment died in 1997 no one has broken the 120 barrier. It is quite possible – even probable – that, barring some fairly profound medical advances, I will not live to see someone break her record.

Why? What sets this upper limit and what purpose does it serve for a species to have a fixed maximum lifespan? To understand this it seems we need to understand what ageing is and how it occurs.

Ageing is a process shrouded in misconceptions. Many people seem to think that, like machines, we simply wear out. As to the deeper 'Why?', answers usually involve stuff like 'for the good of the species' and 'to make way for the young'.

It would be tempting to think of the human or animal body like this, but it would be wrong. Yes, bits of us wear out as we get older, just as bits of cars wear out as they get older, but there are fundamental differences between the ways in which the passage of time affects an organism and a machine. For the first few years of our lives it is not clear that we actually 'age' at all. Thanks to our huge brains, human infants have to be born effectively prematurely; otherwise their mothers simply would not be able to deliver them. Compared to the young of virtually all other mammalian species, human babies are extraordinarily helpless. For the first few years of our lives, time tempers, not ravages, our bodies. Bones grow thicker and stronger, as do muscles, tendons and cartilage. Our brains become sharper, we discover the joys of locomotion, solid food and language, and we develop a finely tuned immune system. For the first 20 years or so of our lives we actually 'anti-age' – in real terms, from babyhood to our late teens we get, effectively, younger – our muscles get stronger, our skeleton strengthens, our immune systems become more powerful, we become sexually mature, our skin tightens and our hair becomes thick and lustrous.

Sometime in our late teens or early 20s, things start to go downhill. Ageing, or senescence, is loosely defined by a declining ability to deal with stress. The body's repair mechanisms, including resistance to infectious disease, become less efficient. This isn't because 'wearing out' is the inevitable result of the exposure of a living organism to time – see what happens to us in early life – so there must be some other reason for our bodies to give up on keeping us young.

One way of looking at an 'explanation' for ageing, or at least an explanation for why different species age at different rates, is to consider it in an evolutionary context. We age at a rate determined by how long we could expect to survive in the wild. Big animals tend to age slower than small ones simply because, being big, they are less likely to be eaten, killed in an accident, die quickly of starvation, or perish through cold. There is no point in us having evolved an efficient mechanism for dealing with cancers in our 90s, as our ancestors would most likely have been in the stomach of a sabre-toothed tiger by then.

An elephant will live perhaps 20 times longer than a mouse. Mice in turn are much longer-lived than ants. The ageing–size 'rule' cuts across phyla and order. Big reptiles live longer than small ones; ditto fish and amphibians. It is the interesting exceptions to this life expectancy–size correlation that prove the rule. Mice are short-lived, but bats (also small mammals) live a relatively long time. Most birds live far longer than one would expect given their size and weight. Tortoises (even small ones) live a long time, and large tortoises may be among the longest-lived animals of all. This can be explained simply. Creatures that can fly, or which have body armour, are far less likely to be eaten in any given time than creatures which do not. And being eaten is a major – often *the* major – cause of mortality in small animals.

It makes little sense for a mouse to be equipped with (perhaps expensive) repair mechanisms for its body and the DNA within if the chances are that it will be owl food within a year or two. Better throw everything into making the mouse grow up as quickly as possible and reproduce as effectively and rapidly as possible in the short time that will inevitably be available. Elephants are far less likely to be eaten than mice, so it makes sense for their bodies to allow them the luxury of an old age in which they can rear several generations of young. They live slow, and die old. Like us.

The evolutionary theory of ageing leads to the nice aphorism, 'death is the price we pay for sex'. By throwing all its eggs, so to speak, in the basket of reproduction quickly enough to get a chance of passing on its genes, the body pays the price later on in the shape of cancers and other diseases of old age. The idea is that the very sex hormones which endow our genes with immortality hasten the extinguishing of the temporary ark in which they are carried from one generation to the next.

Maybe the evolutionary theory of senescence explains something about why we age but not how – nor what ageing really *is*. Evolutionists surmise that polar bears are white because that is the only colour to be if you want to sneak up on dinner in the Arctic. But this doesn't tell us what *makes* a polar bear white – to answer this more basic (and in this case more straightforward) question, you need to catch one and look at its fur, how the follicles are constructed, how these refract and reflect light and so on. But if you speak to gerontologists you are left with the feeling that there is nothing simple at all about the how of the ageing process; in fact, that there is a big hole in the middle of what we understand about ageing – *what* makes the fur white, not *why* it is white (although they will usually not put it in these terms and indeed will deny, usually, that the hole exists). Is ageing a process, a deep, underlying genetically driven 'clock' that starts ticking the day we are born? This is the 'programmed' theory of ageing. Or is ageing simply the name we give to the gradual build up of cellular and DNA damage, largely from the environment and by oxygen in the body, leading to a cascade of diseases? This, broadly, is the 'stochastic' theory.

Oxygen plays a starring role in the story of senescence. If death is the price we pay for sex, it might also be the price we pay for not being yeast. Some time around 3,000 million years ago, life invented photosynthesis. In the space of fewer than a couple of hundred million years, the air started to fill with a terrible poison. The nice cosy blanket of nitrogen, water vapour and carbon dioxide was suddenly polluted with a noxious gas with a terrible knack for ripping apart complex organic molecules and playing havoc with the genetic code.

Nowadays we tend to think of this gas, oxygen, as a rather good thing. You can do without food for six weeks, or water for six days. You can't last six minutes without a fresh supply of this reactive molecule to go ripping through your body. But oxygen is still a poison, which is not something that many people realize. If you breathe pure O_2 for any length of time your throat will become raw and your airways inflamed by the gaseous equivalent of a sugar-rush.

By adapting to oxygen and, eventually, co-opting it into clever respiratory cycles, life hauled itself out of its anaerobic stupor and started whirling and thrashing about with untamed vigour. Oxygen is powerful stuff, allowing large amounts of energy to be released quickly and efficiently when molecules are broken down. But oxygen is a two-edged sword. Its very reactivity damages the cells, and especially their nuclear material, that allow our bodies to operate so quickly. Just as plodding diesels tend to outlive highly tuned racing car engines, our supercharged, oxygenated metabolisms contain the recipe for their own demise. We start to rust, from the inside out.

But oxygen isn't the whole story. Another suspect in the crime of death is the telomere. Telomeres are segments of DNA that lie on the end of our chromosomes and they play a vital role in preserving genetic integrity during cell division. Every time a cell divides, a bit of its telomeres is lost. One theory is that the inevitable shortening of telomeres is, in effect, the ticking clock of ageing. Once they become too short, the cell loses its integrity. In 1965, the biologist Leonard Hayflick discovered that differentiated cells dividing in culture can only divide about 50 times before dying. This is known as the Hayflick Limit, and the shortening of the telomeres is thought to be the cause. It is also thought that if a way could be found to prevent the shortening of the telomeres then this limit could be circumvented, and considerable research is going on in this area.

But telomeres may be a red herring. It certainly seems to be the case that removing the Hayflick Limit may not be a good thing. The two kinds of cells that can replicate indefinitely are all-purpose stem cells and cancer cells. By interfering with the telomere-shortening process we may simply end up triggering the growth of runaway cancer. And it is not even clear that the shortening of telomeres *causes* ageing, or whether the relationship is the other way round. In some species of seabird, telomeres actually lengthen with age. It is extremely difficult for science to separate out the causes and effects of ageing.

Professor Steven Austad of the University of Texas is one of the world's leading gerontologists and also one of the people best able to explain, simply, what is going on as we get older. The evolutionary theory explains why we age; *how* we age is basically down to the fact that we rust and that we cook. Our cells, or the bits and pieces of organelle and metabolism contained within them, are destroyed by oxygen as we burn fuel. Austad likes the clock analogy, he says. 'What we don't understand is what sets the speed of the clock. Damage happens for a reason and happens faster in a mouse than in a dog and faster in a dog than in a human. The key question is: what is responsible for these differences?'

'Ageing is a process which makes us increasingly vulnerable to various diseases. It is surely something other than the sum of our diseases though. Even the healthiest 50-year-old cannot sprint as fast as he could at 20. That's ageing in its purest form.' As we age, our bodies' ability to cope with various kinds of damage declines. Specifically, the ravages of reactive oxygen and ongoing damage to our DNA seem to be key to this process. But unravelling the nature of cause and effect is proving difficult. If we see changes in our cells as we age, are these changes the cause of the degeneration, or are they the effects of some deeper underlying mechanism?

One interesting consequence, Austad points out, of the ageing-as-process idea is the fact that we all become much more different as we get older. Twenty-year-olds are much more like one another than 80-year-olds, to put it crudely (to put it most crudely, nearly all 20-year-olds will be alive, while a goodly proportion of those potential 80-year-olds will not).

And it is true. We all know of people in their 80s and even 90s who are quite capable of leading the same sort of active lives that most take for granted in middle age or even youth. I know one mid-septuagenarian who still goes hiking in the Alps regularly, while I know plenty of people in their 50s or even 40s unable to run a mile. Much of this is down to lifestyle, to things like smoking and exercise and a lifetime of healthy eating and so on, but much of it, even the bossiest of health experts agree, is down to luck.

However lucky or virtuous you are, you will still grow old and die. The question is, can anything be done about this? In a way, it is surprising that more biologists are not working on this problem. After all, their lives depend on it.

Gerontologists, people who study the biology of ageing, disagree with each other, like all scientists. But there is one thing they tend to agree upon, and that is that ageing is not something we should try to stop. Leonard Hayflick has roundly condemned the life-extortionists, people who want to use science to extend human lifespan. It is rare to meet a gerontologist who thinks that we should be trying to do this, even if it is possible. Ring up just about any biologist in their field and they will say the same thing. 'We don't understand everything about ageing. But what we do understand indicates that slowing the process down will be hugely difficult and expensive.' And, this is the key bit: even if we could, we shouldn't.

Why not? Here science drifts into morality. It seems to have become a mainstream liberal view that altering human lifespan is somehow wrong. One argument is that by making people live longer science will contribute to the overpopulation problem. Another is that research into anti-ageing therapies will, as soon as it shows any sign of success, inevitably siphon off just about every research dollar and euro going, to the detriment of the rest of medicine. Finally, it is probable that any 'cure' for old age will have to be rationed by cost. There will be yet another divide between the world's haves and have-nots.

These are rational arguments, but I believe there is another unstated reason why so many scientists are opposed to anti-ageing research. This is something akin to scientific Puritanism, the feeling that whatever the logical objections and justifications might be, this sort of thing just isn't *right*. This is strange, as there have been few objections, if any, to the great strides that have been made in the past 200 years when it comes to life *expectancy*. No one objected to vaccination or antibiotics on overpopulation grounds, and although there is an extremely vexed political issue concerning the cost of antiretroviral drugs and their subsequent availability to the poor, no one is suggesting that the research into a cure for Aids should stop. This puritanical feeling is less strong than it was, but it is still there.

So far of course, the argument is academic. Despite the money spent, the research undertaken, the promises and the marketing campaigns, the fact remains that not a single potion, pill or treatment has succeeded in extending human lifespan by one day. That is not to say that some people are not trying. There are the beginnings of strategies out there which may, one day, yield tangible results. The era of the 1000-year-old human may be a long, long way off, but the era of the 150-year-old might not be.

Perhaps the simplest way of extending human lifespan might be to make it illegal to have children before the age of 40. Michael Rose, professor of evolutionary biology at the University of California, is one of the few biologists I have spoken to who is enthusiastic about increasing human lifespan. He famously bred 'Methuselah' fruit flies that lived some 10% longer than average by using a form of simple artificial selection – only allowing the oldest animals to breed. 'Oh yes, if you were a dictator you could increase human life expectancy quite dramatically, simply by making it against the law to have children before you were 40 or 45', he told me. The reason this works is quite simple – you are removing from the gene pool anyone unfit enough to still be healthy and fertile into middle age.

This won't happen. What may happen is some sort of gene manipulation. A number of 'ageing genes' have been identified in organisms such as the fruit fly (*Drosophila melanogaster*) and the nematode worm *Caenorhabditis elegans*. Extra copies of a gene called Sir2 have been found to extend the lifespan of both organisms. One of the most interesting discoveries in recent years is how many key genes are shared right across the range of species, from yeast to humans. One gene, called RAS1, is linked to yeast lifespan. We have a version of this gene, although it is not yet known if messing around with our version of this gene will have the same effect.

A strategy that seems to have great promise derives from one of the most depressing findings in the whole canon of medical science, namely that severe calorie restriction seems to prolong lifespan. There is plenty of evidence, from rats mostly, that keeping animals in a state of near starvation (while maintaining a balanced diet rich in vitamins and minerals and so on) has a definitely beneficial effect on lifespan.

In studies going back more than 70 years, putting rodents on brutally spartan diets has been demonstrated to extend their lifespans by as much as 50% – the equivalent of getting humans to live into their 150s or so. The calorie restriction (CR) effect seems to be no respecter of species or even phyla: worms, spiders and insects have all been shown to benefit. The first proper human experiment in CR took place in 1991 when eight scientists sealed themselves inside a series of linked geodesic glass domes in the Arizona desert. The project was called 'Biosphere 2', an attempt to create a self-sustaining habitat completely hermetically sealed from the outside world, 'Biosphere 1'. One of those scientists was our Roy Walford. He concocted an iron diet of a thousand or so calories a day, containing all essential nutrients.

When the scientists emerged from their dome a couple of years later the only one not in good shape was Walford, who had hurt his back terribly in a fall from a girder. In subsequent years, Walford promoted his life extension theories based on low-calorie diets. The problem is that humans live a very long time, and getting experimental data on lifespan extension in humans is extremely difficult (it is much easier with fruit flies and rats). The idea that to guarantee a ripe and healthy old age we need to spend our lives in a state of near starvation in one of the most depressing in the history of medicine. Could there be other ways?

It is unclear at present how and why CR 'works'. A popular interpretation is that with less food there is less metabolic activity and therefore fewer free radicals released into the bloodstream by the cells' mitochondria, which are forced to work at maximum efficiency. Free radicals and other oxidizing chemicals are thought to perform a key role in the ageing process. By-products of our hunger for oxygen, they literally rust our bodies away from within. A rival theory, the 'Hormesis Hypothesis' states that being in a state of near starvation stresses the body, generating a response not unlike the immune response triggered by infection. This 'hardens' the organism, placing it in a defensive mode and making it overall better able to fight the diseases of ageing. Interestingly, the same Sir2 gene linked to ageing rates in flies and worms seems also to be affected by caloric restriction. This opens up the possibility that it may be possible to mimic the effect of a starvation diet using drugs or gene therapy.

Anti-ageing research is slowly becoming more respectable. Most scientists still seem to be against it, but there is a shift in attitude, particularly in America. Partly this is a function of increasing wealth. Countries like the US are now far richer, in absolute and relative terms, than any societies in history. A poor man in today's America or Western Europe has access to riches beyond the wildest dreams of the Roman Emperors. The rich today, and there are a lot of them, can literally buy almost anything they want. Planes, cars, large chunks of whole countries. But they cannot, as yet, buy more life.

Some scientists, Greg Stock among them, think that to break the magical 125-year barrier we will need to change the human germline, to alter our DNA. It may be possible to use a combination of gene therapies and other drugs to mimic the effects of calorie restriction without being hungry and tired all the time.

Aubrey de Grey, an elaborately bearded scientist at Cambridge University attracts almost universal derision among the 'ageing community' for his thesis that a 'solution' to old age is just around the corner, probably consisting of a cocktail of drugs and genetic therapies that will counter the effects of free radicals and other harmful metabolic processes that weaken the bones, make our skin brittle and cause our organs to slowly fail. The attacks on de Grey are motivated by rational scepticism, but also, I think, by that puritan morality that states that life extension is a place where science has no place going.

It is very probable that all the great advances in extending human lifespan have already happened. Even when infant mortality is taken out of the equation we are still in far better shape than our great grandparents. A healthy 60-year-old today can expect to live maybe another 25–30 years, and that is the first time that has been the case in human history. Life expectancy is continuing to rise in almost all societies, although it remains to be seen whether the current obesity epidemic, prevalent in many Western countries since the late

1970s, will affect this trend in years to come (so far, strangely, it seems not to be doing so). Immortality is a long, long way off.

We can make rats live a long time by starving them, and we can increase the lifespan of worms by tweaking their genes and of flies by simply delaying the breeding age. But humans are not mice, flies or worms. The man or woman who cracks the ageing problem will become one of the most famous people in history. I certainly wouldn't bet on it never happening. But sadly I will almost certainly not be alive to see it.

4

what are we going to do with the stupid?

We don't laugh at people's DNA any more. Or refuse to give people with certain genes, but not others, jobs. Nor do we enslave them or prevent them from voting. And if we do, it is generally accepted that we are doing a Very Bad Thing indeed. And rightly so. This is new, of course. Just two generations ago it was perfectly acceptable to mock and discriminate against those of different skin colour and ethnicity, the very small and the very tall, the brawny and the weak. in my country, people, even quite nice people, talked openly about 'Jewboys' and 'darkies', 'shortarses' and worse. Those condemned (perhaps) by their DNA to be homosexual were not only discriminated against by the *hoi polloi*, but by the law.

It seems quite extraordinary now that in my lifetime it was actually illegal for grown men in Britain to go to bed with each other and have sex, and that hundreds ended up blackmailed, ruined or in gaol for doing so – right into the not-so-swinging Sixties. You could be discriminated against too for plenty of non-genetic stuff. The one-legged, the crippled, the deformed and the burnt were all subjected to often horrible mockery. The past was a pretty ghastly place.

Happily that is all gone now, in polite countries at least. (Crude genetic discrimination is still alive and kicking in much of the world. There are countries, for example, where the possession of testicles is deemed necessary to pilot an automobile.) It has been swept away by rafts of equal opportunities legislation, a new, liberal right-thinking consensus that operates more or less across all mainstream political boundaries. The current leader of the British Conservative Party, traditionally the home of socially robust attitudes, has no detectable homophobic or racist bones in his body. The 'differently abled' are helped and ramped and made space for. Laughing at someone because they are black or Jewish now seems almost quaint, like poking fun at the peasants or employing court eunuchs.

But there is one area of genetic prejudice that remains perfectly acceptable at all levels of polite society and, indeed, has arguably become more acceptable in recent years than ever before, and that is the prejudice against people born with a low IQ. 'What to do with the stupid', to put the problem at its crudest, is perhaps the greatest socio-political problem of our time. It is a problem fuelled by scientific ignorance, wilful misunderstandings and extreme prejudice. It is something that scientists and social scientists are squeamish of even discussing – just like life extension in this respect. Both left and right have come up with solutions – solutions which, as we shall see, are equally unsatisfactory.

Why on earth is a question like this in a book about scientific mysteries? Surely this is politics, political correctness, the stuff of education debates and sociology? Well, yes. But the issue of variation in human intelligence goes right to the nub of one of the most venerable and vexed scientific questions of all: the relative importance of genes and environment in how an organism develops during its lifetime. It is no mystery that some people are brighter than others (although some people still seem to dispute this). But it still seems to be deeply mysterious as to the best way of going about dealing with this. Our social scientists and economists have chosen, and for the most part still choose, to ignore this problem.

There are some barriers to overcome when talking about human IQ. For some, even to consider this issue is profoundly unacceptable. 'What do you mean, what should we do with the stupid?' a friend asked me after I wrote an essay on this subject for the *Spectator*, a British political magazine. 'The very asking of that question is going to offend a lot of people.' And of course he was right. Even suggesting there is a problem is going to cause trouble. For a start, debates about the relative intelligence of different people always risk straying into some very dark waters indeed (although of course they shouldn't).

Even talking about IQ gets you into trouble in some quarters. Lots of people decry IQ tests as meaningless – 'they are a measure of your ability to pass IQ tests' is a common sneer – but the fact is that while a good test may not be measuring pure intelligence *per se*, it does seem to be measuring something that correlates very well with it. People who score highly in IQ tests tend (not always of course) to be generally regarded as 'bright'. People who score very highly indeed may well be a bit strange, eccentric, weird and lacking in social skills, but many more are not. Life's achievers are predominantly drawn from the ranks of the high-IQed, not low. Let's assume for the sake of this argument that the term IQ does mean something.

So who are we talking about when we use the word 'stupid'? We are not talking about the profoundly mentally handicapped, nor those who have lost significant cognitive function through ageing, injury or disease. In most wealthy and humane societies people with – to put it very crudely – an IQ much below 70 are cared for by the state, or with some state help and guidance, either in or out of the community. These people are often not considered 'ill', yet it is also accepted that they will find it hard to take an active role in society without at least some outside help. No, the people I am talking about fall into an unhappy no-man's land on the IQ scale – too bright to be regarded as truly disabled, but for the most part lacking the intellectual skills to survive easily in what is becoming an increasingly knowledge-based world. These people suffer discrimination at every turn, discrimination wired into the whole fabric of most societies on Earth. And it is discrimination that is getting worse.

Something that is not appreciated at all is the fact that it is quite hard for a person living in a technological society, the 'modern world' if you like, to earn a living wage if he or she is illiterate – something that was certainly not the case say 150 years ago. Then there were plenty of jobs that relied purely on brawn rather than brain. In 1900 more than a million British men worked in the mines, and millions more in the fields. During the 19th century millions of immigrants poured into the US, some of whom became doctors and lawyers, but the vast majority went to work on farms and in the new factories. Many of these blue-collar jobs were highly skilled and required training and intelligence, but most were not. Now most of those jobs have gone.

This is not of course to say that all those old miners and farm workers were illiterate or of low intelligence. Many were extremely bright. In fact, in the old days, when one's place in society and one's job prospects were determined largely by accidents of birth and geography, you could probably find pretty much the same distribution of mental abilities among the 'lower' classes as among the wealthy. This is something that cannot be true today.

The proportion of the workforce who can carry out their paid tasks with no educational qualifications is far smaller than it ever has been. To earn a decent working wage in most Western countries you need far more brain than brawn. Of course, there are still jobs open to those with purely physical skills, but they are far fewer in number than ever before. Most of the labourers are no longer needed, and unskilled manual work is increasingly being farmed out to poorer countries. Only a tiny elite can earn a living playing sport. You may not have to be that bright to be a successful footballer or model, but you do need a highly unusual physical talent and/or physique.

In Western countries, jobs considered 'blue collar' or 'working class' now often demand significant intellectual input. The modern 'factories' are the call centres, giant warehouses full of unfortunates making and answering endless telephone calls concerning bank accounts, insurance and the whole panoply of life that cannot be dealt with in a retail store or on the

Internet. Many people mock these jobs and those who do them, forgetting perhaps that they require significant skills – a good command of the language and of the nature of the business to hand, good conversational skills and the ability to operate a complex computer system at some speed. Working in a call centre is not the equivalent of baling hay or breaking rocks.

So what are the options open to the people who simply cannot join in modern life? Precious few, it seems. One function that the less bright can perform is to act as society's collective court jester. Think of all those hideous and cruel TV shows designed to showcase the dim for our delectation (for that is the real point of *Big Brother* and *Wife Swap* and *Holidays from Hell* and all those dreadful 'reality' programmes that appear to have got the world hooked). Of course, we have always mocked our fools – but it seems to me that the vitriol directed at the not-very-bright is a nastier and more powerful concoction today than it has ever been. And as the old class barriers break down and the tectonic plates of society realign, it is these people who sink inexorably to the bottom. And, not being bright, they lack the wherewithal – and especially the leadership – to get themselves out.

The dangers of a society where intelligence and effort are rewarded more than accidents of birth were first noted by the social scientist Michael Young. In his 1958 satire *The Rise of the Meritocracy*, Young pointed out that in a society where your status was defined by your merit – he defined this as 'intelligence + effort' – the elite will tend to feel wholly entitled to the privileges they enjoy.

This rather unexpectedly contrasts with the old class-based system, which, for all its raging, foam-flecked snobbery, tacitly acknowledged that your station in life was pretty much a matter of luck, and hence one should not brag too much about one's success or sneer at those less fortunate than oneself. The new 'meritocratic' way of thinking can now be seen in the United States (although in fact America is not nearly as meritocratic as it likes to think). In the US, people talk freely about 'losers' and 'folk who take the bus' in a way that seems extremely alien and unkind to people from older, perhaps more traditionally stratified, societies. Failures are failures because it's their fault. But, of course, when it comes to intelligence it isn't, any more than it is someone's fault for being black or Jewish, or very tall or blonde.

So who are these unfortunate people? In most developed countries, roughly 68 per cent of the population have an IQ between 85 and 115, and the numbers living above and below these limits are roughly equal.

Within and above the 'average' range, most people will be able to function fairly well in our society, but what about the 10–15 per cent or so that lie below IQ 85 and above 70? Today many of them inevitably become the genetic underclass of IQ-challenged unemployables, forever drifting along the crime-driven, drug-using margins of our society. Lots of them are in prison (along with the mad, the sad and the bad). Being working class was a rum deal; being underclass is far, far worse.

Intelligent people, like tall men and pretty women, have always earned far more money, had higher status and happier marriages, been healthier and less likely to be in jail than the stupid. But the point is that this is becoming more the case as society becomes more 'advanced', and driven by technology and data. One interesting corollary of the value we now place on brainpower is the relatively new phenomenon of the desirable alpha female.

Being extremely bright, if you were a woman, was not much of an advantage in most societies until quite recently. In Victorian Britain, for example, the opportunities for a woman to earn her living through brainpower alone were extremely limited. According to the 1901 Census, there were fewer than two hundred registered female physicians in the whole of the UK. Going to university was difficult and expensive – most did not even allow girls to study. You could become a teacher, perhaps, or a governess, or maybe, if you were exceptionally talented, earn your living writing or in the arts. Most of the professions were closed, as was the world of business.

Being bright and female did not even increase your chances of finding (or being found by) a desirable mate. In fact, it lessened it, as many bright women were loath to make the financial sacrifice that would inevitably follow marriage. Traditionally, in Europe and its offshoots at least, alpha males, in having the pick of the females, would tend to choose on looks and background rather than intelligence. The dumb blonde hanging off the arm of the successful politician or businessman is a horrible cliché, but it has an element of truth.

Things are changing. Alpha females use their brainpower to their economic advantage, and are thus now far more desirable mates. Powerful, successful men these days seem to be choosing equally powerful, successful women as partners in a way that did not happen in the past. This is having some interesting effects according to some scientists.

It may have a polarizing effect on the IQ 'bell curve', the statistical graph of intelligence quotients across the population. With the brightest men and women being effectively taken by each other, the overall effect may be a flattening of the 'bell' – more very bright people, and perhaps more very dim. It is too soon for this effect to be seen; it is, after all, only very, very recently that the whole phenomenon of the alpha female has existed at all. As late as the 1960s the marriage bar still existed in most professions, and women in government jobs were, quite openly and officially, paid less than their male counterparts for doing the same job. There is still a 30% salary discrepancy, although the reasons are now less clearly stated.

Another interesting suggestion is that these changed mating patterns may lie at least partly behind the puzzling epidemic of autism. Higher – far higher – percentages of children are now diagnosed with 'autism spectrum' disorders than were even two generations ago. The figure for 1970 in the US, for example, was 1 in 2500; today it is 1%, an astonishing increase mirrored in many other countries. The cause is a mystery. It is too recent to be some sort of general genetic shift. Environmental factors have been mooted, particularly exposure to certain vaccine combinations. Many people in Britain remain convinced, after high-profile media campaigns, that the MMR triple vaccination routinely given to young children can trigger autism, even though the scientific consensus is that there is no such link. Of course, much of the increase in autism can be attributed to better diagnosis. But this cannot explain all of it.

One plausible explanation has been offered by Simon Baron Cohen, a psychologist and director of the Autism Centre at Cambridge University. He thinks that 'assortative mating' may play a part. Specifically, it is becoming more likely that men and women with 'male brains' – logical and systematic – will mate with each other and have children. In a 2006 article in *Seed* magazine he wrote:

> Consider that in the late 1950s, less than 2 percent of undergraduates at MIT (a university that caters to people with good systemising skills) were women. Today female enrolment has jumped to 50 percent. This microcosm is just one example of how society has changed in ways that would bring strong systemisers into greater proximity.
>
> Over the same period, air travel has also meant far greater opportunities for people from widely differing backgrounds to meet, possibly brought together by their common interest in systems. Finally, over this same time-frame, individuals who are systemisers have enjoyed new

employment opportunities as the result of the digital revolution. Where 50 years ago a strong systemiser might have found a job as an accountant, today every workplace needs computer-savvy employees, and the financial rewards for good systemising skills can be immense.

Baron Cohen's thesis has attracted some controversy, both among those who are convinced that autism must have a primarily environmental cause and among others who claim that talking about any aspect of mind and genes in the same breath is tantamount to being a Nazi.

To the blank-slaters, those ideologically opposed to any innate differences in mental ability at all, the traditional satisfactory solution to the IQ problem is to deny that it exists. This sort of thinking led to the bizarre and hugely counter-productive social science model, which took hold among the educational establishments in the West in the 1960s.

This is the traditional solution of the well-meaning left, to pretend that the stupid simply aren't there. This is dangerous, and harmful, mostly to the stupid themselves. Educational policy in many Western countries, most notably Britain and some parts of the US, still reflects this thinking. By pretending that all children are equal, and having a bitter insistence on inclusivity (which can also see the quite severely disabled educated in mainstream schools, to the detriment both of them and of their more able classmates) the less bright have been left floundering, failed by their teachers who are forced to teach that mythical average child (the one who often, mysteriously, fails to show up for classes) and by an examination system that rewards only academic merit at the end. Thus millions of teenagers leave school branded as failures, unable to meet the gold standards academically and yet with no alternative training in skills in which they might be able to excel.

Much more is made of the fact that this system also fails the very bright. It can, but I would argue to a lesser degree. The bright will usually look after themselves and tend anyway to receive most of their basic education from their parents and from books. One's advantages in life tend, after all, to be won before one is even born. No, it is those who are below average who suffer most when it is pretended that everyone is the same.

Believing that the stupid are not there is, on the face of it, an extraordinary thing to believe. For to do so we must accept that human (or any other) intelligence has little or no genetic component. Now, unless our minds are the property of some nebulous ghost in the machine, and not of flesh and blood, this surely cannot and need not be so. I have never understood why people who are quite happy to accept genetic differences in, say, athletic ability, susceptibility to sunburn, and even complex traits like sexuality, are still fundamentally unable to see that our minds may have a strong genetic component too.

Obviously and facetiously this must be so, otherwise humans would be no brighter than halibuts.

In British author Matt Ridley's excellent book, *Nature via Nurture*, the way in which environment and genes work together is explained. If you have Einstein-class DNA, you still need supportive parents, and books and schooling and so on, for your potential to develop. Your environment unlocks the potential in your genes (just as a good diet unlocks your genetic potential to be tall). But that doesn't mean that your genes don't matter. The left's solution – there are no stupid people – is simply not viable.

So what about the right? What do they have to say? By the 'right' I include those Communist regimes whose educational systems were largely based on the rigorously streamed Prussian model that was copied in the old Soviet Union. I also include those counter-revolutionary educationalists who clothe themselves in traditional rigour, and also self-avowedly meritocratic societies and institutions which consciously reward 'ability' when that ability always has a large intellectual component. The right's solution to the stupid is to accept that they exist but then to go on and ignore them and wish they would go away.

It is interesting to speculate about what will happen to human intelligence, for there are strong and mutually contradictory forces that appear to be at work here. A hundred years ago brawn may have mattered far more than it does now, but a hundred thousand years ago the opposite may have been the case. We do not know what drove the evolution of our bizarre, energy-hungry brains, but it is thought that the ability to develop and acquire technology was probably a happy by-product of the process rather than the primary cause. A convincing argument has been made for our intelligence developing as a social implement. We have big brains because they are good for gossip. Did the acquisition of language precede the acquisition of intelligence? We don't know.

Whatever its origins, the big human brain became a useful survival tool. High intelligence would have granted perhaps greater social status and therefore greater choice of mate. Being clever would also have improved survival prospects, especially as technology developed. Clever people would have been better hunters and cannier gatherers, and probably better at looking after children and making plans. Before the development of specialized labour, which came along with the neolithic farming revolution, a successful human probably had to be something of an all-rounder. This is pure and idle speculation, but it may well have been the case that pure brawn and brawn alone was no more useful in the Old Stone Age than it is now.

We know that human brains grew over time. The skulls of people who lived 80,000 years ago were always bigger than those of the hominids who lived at earlier times. High intelligence confers such obvious advantages that it is most probable that the driver for this brain expansion was simple natural selection. But that raises the question: what will happen now? Some scientists talk about *Homo sapiens* as being the first post-evolutionary species. With a conscious awareness of our own evolution, plus the technology and wherewithal to influence and even negate the traditional forces of natural selection, what will happen to our brains now? Will they shrink? Or carry on growing?

One argument says that we will become dumber. In most Western societies, the argument goes, those of low educational attainment tend to have more children than those with university degrees. Since intelligence has a high inheritable component, this will tend to drive average IQ down. But there are problems with this argument. Firstly, the tendency of the lower socioeconomic groups to have more children than other groups is rather new and geographically limited, and may be very transient. Traditionally, in Europe say, it was the wealthy who could afford to have more children. Upper class women would traditionally marry earlier and have more babies than their poorer sisters, as they would have been depended upon to make an economic contribution by working. It is true that in countries with a strong welfare state, fecundity is quite strongly correlated with low socioeconomic status, but this is not universal by any means. In countries where 'welfarism' is less prevalent, such as the US, the wealthy and educated can and do outbreed the poor.

Secondly, there is the undeniable and rather mysterious fact that we seem to be becoming brighter. In just about every industrial society, average IQs have risen quite dramatically in the last half a century or so. This cannot be evolutionary – there have not been enough generations for natural selection, genetic drift or some other evolutionary mechanisms to have taken effect. Something must be acting on the raw genetic component of intelligence to bring out the 'best' in our brains. What was 'quite bright' three generations ago is now merely average. What could it be that has caused this?

There are several possible candidates. Better diet is one. This could explain the dramatic rise in IQ levels seen in some Asian countries since the Second World War. But this would not adequately explain why IQs have also risen in countries like the US, UK and France, where diets have not changed significantly in 50 years. Indeed, the British probably ate better during and just after the Second World War than at any time before or since. Food rationing meant more, and better quality, food for a majority of the population. Yet austerity Britain was probably a bit dimmer than it is today.

Better education must have had an effect, but again there is little evidence that this could have had much of an influence in the short time-scales involved, in Western European countries at least. A popular, but left-field, suggestion is that it is all down to television. Far from being an idiot-box, the TV may be literally pummelling your child's brain into shape, producing a welter of complex non-stop visual stimulation. According to this theory, it may not even matter what sort of television you watch; cartoons are probably as good as the Discovery Channel. It's a nice idea (computer games may play a part too), but it is wholly unproven.

The low-IQ problem will probably never be tackled. Intelligence, unlike skin colour, athleticism, height, weight or any other gene-influenced attribute goes to the very heart of what it is to be human. The less-than-bright are seen as less-than-human. They are being left behind faster and faster, and the phenomenon will become global if and when the developing world starts to catch up with the industrial countries economically. In the most fiercely technocratic societies, like those in some Asian countries, where success in life is determined almost entirely by the ability to pass a series of extremely tough academic hurdles, being dim must now be very, very hard indeed. One answer of course is to massively improve the quality of education available. A good proportion of any society's intellectual also-rans can be brought into the IQ mainstream with better schooling. But with the best will, and the best schools, in the world, there will always be some left behind.

The not-very-bright are mocked, are less healthy, have a smaller pool of mates from which to choose, are more likely to be unemployed and poor, and are far more likely to drift into a life of crime. They exist in a society which not only derides them but which places a series of hurdles in their way deliberately designed so that they will fail. This is manifestly unfair. When is somebody going to do something about it?

5
what is the dark side?

In late 2005 I was privileged to gaze upon one of the marvels of the modern world. Buried around 30 metres under the Swiss–French border, near Geneva, is a roughly circular tunnel, about 27 kilometres long. The tunnel is lined in grey concrete, with a painted floor, and is about three metres across. It makes a nice, if somewhat dispiriting, running track, the wall curving endlessly and rather hellishly away in both directions as you puff your way around. In fact, informal races have been held in it; a good time for a circuit is two hours. A more common means of transportation would be a bicycle or one of those two-wheeled electric buggies. I was lucky to get into this tunnel, because from now on and for a good while no one else will be going down there.

The tunnel forms part of one of the most impressive engineering feats in history, part of a scientific instrument gargantuan in both scale and intent. Like the greatest of the telescopes, atop their Canarian and Chilean peaks, the big colliding tunnel at the *Conseil Européen pour la Recherche Nucléaire* (CERN), is a multi-billion-dollar behemoth designed to expose the deepest realities in our Universe.

At several points around the CERN ring are some of the most awe-inspiring bits of engineering I have ever seen: vast, unfathomable polygonal rings, made of Russian pig-iron and British steel, French titanium and German plastic, and several million metres of wiring and ducting. These assemblages are as big and heavy as ships, and have been lowered into artificial caverns as voluminous as cathedrals. This is Big Science. In November 2007, after some ceremony with the on-button, these Brobdingnagian structures will witness the first deliberate collision of packets of particles. The heavyweights of the sub-atomic zoo, protons and neutrons, will slam into each other at near light velocities after being accelerated around the ring by giant superconducting magnets.

The Large Hadron Collider (LHC) is the largest and most expensive physics experiment in history. It has a series of goals, none of them modest. One is to find the elusive Higgs Particle, the object (or should we call it a field?) that is thought to permeate the Universe and to give everything its mass.

And another is to discover the nature of dark matter. It is a good thing that this huge machine has been built, because dark matter is one of the biggest embarrassments in science and its nature one of the biggest unsolved mysteries.

Unlike some of the other problems in this book, dark matter is something of a blank slate; we really do not have an inkling of what it is or where it came from. It is as mysterious as human consciousness. About 4% of the total 'stuff' (mass + energy) in the Universe is thought to be composed of well-behaved ordinary matter, the stuff of stars and planets, you and me. Dark matter, whatever it turns out to be, accounts for another 22% (so there is five and a half times as much invisible matter in the Universe as there is visible matter). The rest is made up of dark matter's even madder and more mysterious cousin, dark energy, of which more later.

We have some ideas as to the nature of dark matter, but nothing concrete. And thanks to the construction of vast atom-smashers like the LHC we may find ourselves going from a position of total ignorance to total understanding about one of the biggest mysteries in the Universe in a matter of just a few years or even months. Now that would be an extraordinary result.

Dark matter is by far the most common 'stuff' in the Universe, and yet no one has really the faintest clue what it is. It is hoped that the very high energies generated by the collisions in the LHC will blast the particles that form dark matter out of hiding. Whether we will ever be able to get to grips with dark energy remains to be seen.

Dark matter has been bothering cosmologists for some time. Since the 1930s, astronomers have been uncomfortably aware that the amount of matter visible in the cosmos is not nearly enough to account for the movement of the stars and galaxies that we observe.

In 1933, the Swiss astronomer Fritz Zwicky, working at the California Institute of Technology (Caltech), analyzed the movement of a cluster of galaxies and discovered that it could only be accounted for if a large amount of unseen mass – far greater than the visible stuff, which is mostly stars – was lurking in the area. This anomaly has now been seen everywhere we look. The large, visible objects in the Universe (the galaxies and the stars in them) behave as if huge amounts of invisible matter are pulling them about.

So the search for the Universe's missing matter has become one of the biggest stories in science. Theories have abounded. Maybe the dark matter is just ordinary stuff that is hard to see – rocks, asteroids, lone planets and brown dwarf stars too dim for their light to be detected. Or maybe it is made of vast clouds of gas and dust.

Some of the missing matter *is* undoubtedly just this, but most of it cannot be. This much ordinary matter would leave an unmistakable mark in the form of re-radiation of electro-magnetic energy, which we simply do not observe. No, most physicists now believe that dark matter is composed of some novel subatomic particle. A current front-runner is a hypothetical particle called the axion. Proposed in 1977 to clean up a few equations involving the strong nuclear force (one of the fundamental forces of nature), axions make good candidates for the dark matter particle because they have very little mass (although they certainly have some) and will barely interact with matter.

In July 2006, tantalizing evidence for the existence of something very like axions came from the National Laboratories of Legarno, in Italy, where a particle accelerator generated a slight shift in the polarization of a laser beam fired through a magnetic field. The finding is controversial, but in early 2007, *New Scientist* reported that an attempt was being made to persuade the people in charge of a soon-to-be-defunct accelerator called HERA to be pressed into service to try to duplicate the results, something which this accelerator is currently in a unique position to be able to do. But for the moment, dark matter remains, in physics terms, *terra incognita*.

There is good news though. We may not really know what dark matter *is* yet, but at least we have spotted where it lurks. In early 2007, Richard Massey, of Caltech, and colleagues, published in *Nature*[1] the first 'picture' of the distribution of dark matter in a large chunk of the nearby universe. Because we cannot see dark matter directly, we infer its existence from its effect on things we *can* see.

Massey's team made their map from nearly 1000 hours of observation by the Hubble Space Telescope of around half a million galaxies using a technique called weak gravity lensing. Dark matter, though invisible and transparent, has gravity, and thus, like any object, affects light or other radiation that passes through it or nearby. Simply put, if there is dark matter between you and a galaxy you are looking at, the direction of the light from that galaxy will be bent very slightly because the dark matter acts as a giant lens. Massey's team mapped the distortion seen in the Hubble images to work out where these invisible lenses lie.

Writing rather poetically in *Nature*, physicist Eric Linder, of the Lawrence Berkeley National Laboratory, compares this to an old technique used here on Earth:

In the manner akin to Polynesian seafarers who sense islands out of their sight through the deflected direction of

ocean waves, cosmologists can map a concentration of
the Universe's unseen mass through the gravitational
deflection of light coming from sources behind it.

Apparently, dark matter is not arranged randomly. It is
clumped into vast blobs on a grander scale than even the galaxies themselves. Combining the Hubble data with ground-
based observations produced a crude 3D map of dark matter.
It appears to form the scaffold upon which the visible matter
of the Universe, the galaxies, is assembled. Visible galaxy clus-
ters seem to be embedded in vast clumps of dark matter connected by titanic bridges of dark matter called filaments.

Cosmologists now suspect that early on in the formation of
the Universe the dark matter formed a framework around
which 'normal' matter could coagulate. Today, dark matter
also seems to be clumped into ghost galaxies. Recent observations of 'dwarf spheroidals', small dim galaxies which have
been detected orbiting both the Milky Way and the great
Andromeda spiral, show them to consist almost entirely of
dark matter, their ordinary matter stripped away billions of
years ago by the gravity of their giant neighbours.

Edwin Hubble's realization in the 1920s that the Universe is
hugely vaster than the single galaxy that it was once thought to
be, that it was expanding at a furious rate and that the edges of
space and time are likely to be forever beyond our gaze, was
perhaps the ultimate triumph of Copernicanism, the relegation
of humans and their affairs to the periphery of creation. But the
discovery of dark matter and, as we shall see, dark energy,
pushes the 'world of us' even further towards the edge of what
appears to be important. Not only does the little Earth go
round the Sun, not only is our galaxy just one of billions, but it
turns out that even those countless galaxies are made of some
sort of peripheral stuff that is a mere adjunct to the bulk of the
Universe.

If dark matter is *terra incognita*, then dark energy amounts to the dragons that live there. It accounts for the remaining three quarters of the missing mass-energy account of the Universe. It sounds like something out of science fiction, vintage Asimov perhaps, yet it seems to be quite real.

Dark energy is a strange force field which permeates all of space, creating a repulsive force that seems to be causing the Universe to expand. It was first mooted by Einstein. He considered it an unfortunate bodge and his 'greatest mistake'. Today, physicists agree that something very like Einstein's bodge, a repulsive force which stops the galaxies falling into each other due to their mutual gravitational attraction, is needed to explain the Universe that we can see.

It may have been Einstein who first hinted at the existence of a mysterious, all-pervasive energy field, but it wasn't until 1998 that astronomers realized that the expansion of the Universe – actually the expansion of the Universe's fabric, space–time, in which the superclusters of galaxies are embedded – was slower in the past than it is today and is hence accelerating.

The discovery was made by observing distant supernovae, which can be used as markers of the speed of expansion of far-flung bits of the Universe, just as hydrologists throw brightly coloured balls into rivers to measure the velocity of the water in various streams of the channel. One explanation is that gravity works in different ways at different scales. But it is now accepted that some form of energy is pushing the galaxies apart.

In a *Scientific American*[2] article in February 2007, the astronomer Christopher Conselice described the essence of dark energy nicely:

The very pervasiveness of dark energy is what made it so hard to recognize. Dark energy, unlike matter, does not clump in some places more than others; by its very nature it is spread smoothly everywhere.

Certainly, on small scales there isn't very much of it. Every cubic metre of the Universe contains dark energy equivalent (via Einstein's famous equation $E = mc^2$) to some 10^{-26} kilograms, about the same as a handful of hydrogen atoms. All the dark energy in our Solar System amounts to the mass of a small asteroid. But, unlike small asteroids, dark energy is absolutely everywhere. A few atoms' worth in your living room, an atom or so's worth in your head.

And if there is one thing the Universe is not short of, it is bulk volume. Over vast, cosmological distances and vast tracts of cosmological time, the effects of dark energy are, to say the least, substantial. Dark energy acts as a gigantic cosmic sculptor, not only determining the overall expansion rate of the Universe, but also the structure of the smaller-scale 'scaffolding' on which the galaxies are hung.

Dark energy has a negligible effect on anything smaller than the supercluster scale. It is not, for instance, causing our own galaxy to expand (at least not yet). On the scale of thousands of light years and everything smaller, gravity dominates. It is only when you move to the million- and hundred-million-light-year scale that dark energy starts to make its presence felt.

Dark energy could be significant in other ways. The 'current' Universe is quite different to the early one. Today's galaxies, by which we mean the ones relatively close by, are large, stable affairs compared to the often violent, colliding star agglomerations seen when the Universe was, say, half its present age. Star formation seems to have slowed down.

And some of the Universe's most dramatic objects seem to have become extinct; it cannot be just coincidence that the quasars and violent radio galaxies seen at billion-plus light year distances are absent in the nearby Universe. Quasars are thought to be powered by supermassive black holes, superficially, at least, similar to the type found at the heart of the

Milky Way. Yet the heart of our galaxy is (fortunately for us) not a quasar and our black hole is very well-behaved. Some astronomers see the hand of dark energy in this apparent cosmological evolution.

Unfortunately, dark energy does not behave like any other energy field. It is not like gravity (which is generated by mass); nor is it like the nuclear or electromagnetic forces. It seems to come out of nothing, from the vacuum itself. Vacuum energy is a favourite candidate for dark energy, but the trouble is that when physicists work out how much energy should be generated by the random quantum fluctuations of empty space they come up with a value about 10^{120} times larger than dark energy actually appears to have.

This holds the current and rather embarrassing record for a disparity between observed and theoretical values of just about anything in all of science. Dark energy is strongly suspected to be the energy that put the 'Bang' into the big one, and the energy which powered the stupendous expansion of the Universe in the fleeting instants after. But, like dark matter, we really don't know what it is, how it was generated or why, if it was the energy source for the Big Bang and the subsequent cosmic inflation, it has dissipated in power so much since. And while dark energy appears to be a shadow of its former self, some models predict that it will once again become a dominant force in the Universe to rival and even supersede gravity as the major sculpting agent of matter on the scale of planets, or even smaller, rather than on cosmological scales.

One day, dark energy may become so strong that it may rip apart star systems, throwing planets out of their orbits and even shredding the planets and stars themselves. One eschatological scenario has dark energy operating at a level where even individual atoms and bits of atoms are finally ripped apart: the Universe exploding, its death throes dubbed the Big Rip.

In the medium term we (or rather 'we', meaning intelligent life in general, if there is more than just us out there) may have cause to be thankful to dark energy, as it looks certain that if nothing else it will 'save' the Universe from one of its more grisly postulated fates, the so-called Big Crunch. This would be our future if mutual gravitational attraction were to one day overcome the cosmological expansion. But in the end, the dark energy would get us just as surely as if our fate were to be a fiery rerun of the Big Bang in reverse.

Some scientists hope that dark matter and dark energy will simply go away. There are persistent attempts to show that both could be the results of errors in interpretation or substance in quantum physics or relativity. Possibly there is no dark energy and we *will* have to rethink our ideas about how gravity works over huge distances and time-scales. But both dark matter and energy are proving to be stubborn monsters, and neither will vanish helpfully from the equations; the majority view is indeed that the Universe is swarming with trillions of suns' worth of indefinable matter and that the whole shebang is being blown up and expanded by a strange anti-gravity force field that may one day rip it apart.

Dark matter and dark energy are perhaps the Universe's ultimate way of making us feel small. When contemplating the grandest mysteries, humankind has built grand machines. The great henges in southern Britain and France, astronomical calculators, were the Large Hadron Colliders of their day.

It is fitting that the Hubble telescope, very much now in its twilight years, is now being used to map the vast swathes of dark matter which dominate our Universe. Hubble has found it; the LHC at CERN may yet tell us what it is. It is a nice thought that in a few thousand years' time archaeologists may stumble upon the rusting, crumbling remains of the CERN colliders and wonder just what they were used for, just as we do with Stonehenge and its fellows today.

References

1 Massey, R. (2007) Dark matter mapped. *Nature*, Online edition, 7 January.

2 Conselice, C. (2007) The Universe's invisible hand. *Scientific American*, February. http://www.sciam.com/article.cfm?chanID=sa 006&colID=1&articleID=1356B82B-E7F2-99DF-30CA562C33C4F03C.

6
is the universe alive?

If dark matter is odd, life is positively bizarre. Being alive ourselves we take it rather for granted, yet on even the most basic and fundamental levels what it means to be in the world of the quick rather than of the inanimate is oddly unclear and undefined. Life is perhaps the most mysterious property of our Universe. Its existence means that the cosmos is now aware of its own existence in at least one place, and maybe countless others.

We do not know how, where or when life began. We do not know if it started once on Earth or on many occasions. We do not even know for sure if life, *our* life, began on this planet or elsewhere. We do not know if life on Earth is unique, rare or startlingly commonplace. Do we live in an isolated oasis, a cosmos full of amoebae or a *Star Trek*-type universe, teeming with intelligent species? Finally, we do not really even have a proper working definition of what life *is*.

Until quite recently these questions were, rather strangely, not high on science's agenda. Despite the explosion of biological research in the 20th century, the most fundamental questions about life remained strangely on the fringe. As to the origin of life, for example, the consensus view that Charles's Darwin's throwaway line about the first primitive beings arising spontaneously in a 'warm little pond' would more or less do. This 'primordial soup' hypothesis sounded so plausible, so eminently believable, that it was probably thought best to leave it alone.

Speculating about life in space was, until quite recently as well, rather *infra dig*. The whole notion of the alien had been popularized by science fiction, the delusions of the UFO brigade and by people like Percival Lowell and Arthur C. Clarke, who straddled the borderlands between science and fantasy. The fact that when serious space exploration began in the 1960s and it was found that the first worlds we went looking at, the Moon and then Mars, seemed almost certainly lifeless, led to the view that alien life was not really something that proper scientists concerned themselves with.

But it is all rather different now, and the nature and origin of life is one of the hottest questions in science. Astronomers have made several startling discoveries in just the last few years that seem to make the possibility of life elsewhere far more likely than even a couple of decades ago.

'Biogenic' chemicals closely associated with life, from simple elements like carbon to complex organic molecules like amino acids, have been found everywhere we look. We can detect molecules like ethanol spectroscopically in distant dust and gas clouds, and when we crack open meteorites we find far more complex compounds that may be the precursors to life itself. If all that life needs to get going is the right chemistry and a nice warm planet to live on, the inescapable conclusion is that it is everywhere.

?

The origin of life on Earth is still a rather embarrassing mystery. If it were a religious issue the best analogy would be that of the Second Coming. A central tenet of the Christian faith, the return of Jesus Christ, is now more or less ignored by the mainstream churches, despite the fact that in the early days of the church it was a hotly debated issue.

Something quite similar seemed to be the case with biogenesis until fairly recently. Most textbooks seemed to gloss over the problem, much as Darwin himself did, speculating that the horrendously complicated biochemistry necessary for life somehow bootstrapped itself into existence in that warm little pond. Take a handful of carbonaceous gloop (without asking too hard how it got there), add some fire and brim-stone from the nearest volcano, whisk in methane, hydrogen and carbonaceous smog from the primitive atmosphere and round off the recipe by zapping the whole concoction with a blast of lightning and Bob's your protozoan.

This is what we can perhaps call the Frankenstein scenario. Other possibilities for the start of life on Earth can be summa-rized as the Hot Deep Rocks hypothesis, where life is posited to have started either underground or around undersea volcanic vents, rich in nutrients. And then there is the It Came from Outer Space theory. All have their devotees, all have their unarguable points and all have serious flaws. They might all be wrong or, conversely, they might all be right.

Understanding how life started – and how likely it is that we will find it elsewhere – would be a lot easier if we could decide just what it *is*. Living things are made of non-living stuff; there is no 'life element' that always gives a special spark to whatever it is in. We are made from the same stuff as the rocks and the sea, the planet Jupiter and indeed the star Alpha Centauri, none of which, so far as we know, can possibly count as being alive.

So what is it that makes certain jumbles of perfectly normal atoms and molecules alive and others not? Living things, to take a purely reductionist view, can be seen as little more than very elaborate crystals. I remember a schoolteacher of mine, a very dry man who lived in the worlds of mathematics and physics, dismissing the whole field of biology as 'glorified chemistry' and chemistry as 'an unnecessary elaboration of physics'. Biology teachers say that something is 'alive' if it can move, feed, excrete, reproduce and respond to stimuli. But the trouble is, this definition is not really a definition at all, but a contingent description of what we see around us that we have all agreed is alive.

Does life have to be capable of evolution by natural selec-tion? Several definitions of life hold that it does, but why? It is perfectly possible to imagine a biosphere where the dominant mechanism of evolution is not natural selection but genetic drift or some other process. What about artificial organisms? Several people and firms now seem to be a mere grant proposal away from creating a truly synthetic life form.

If we did this and directed its evolution, then would it not be alive? Are viruses alive? They are made of the 'life stuff' DNA and protein, yet they can be crystallized in a petri dish. You cannot crystallize, say, baboons or snakes. Part of the popular view of life is that only life can think, yet clearly most of it cannot, and some things, such as computers, that may be able to 'think' in a limited way (or at least will be able to soon) are clearly not alive.

So, what is life? One very popular if rather floppy definition is 'We don't know exactly but we will hopefully know life when we see it'. Defining what we mean by life on Earth is hard enough, so creating a catch-all definition of life that works everywhere else (assuming that there is life anywhere else) is going to be a great deal harder again.

Scientists and sci-fi writers have, after all, postulated life forms based on all sorts of outlandish chemistries and even more outlandish types of physics. In their charming book about the end of time, *The Five Ages of the Universe*, physicists Fred Adams and Greg Laughlin dream up an intelligent black hole called Bob. He is quite plausible. So are the life forms dreamt up by the gravitational astronomer and novelist Robert Forward, which are made not of carbon, nor even silicon, but neutron matter, the famously bizarre stuff of collapsed stars, a spoonful of which weighs as much as a battleship.

If life depends on organization and complexity (the possible seeds of a definition) then it is quite easy to imagine living beings arising in the magnetized interiors of stars, vast intelligent entities using the quantum property of entanglement, or the huge semi-permanent and constantly evolving whorls and eddies in the atmosphere of a gas giant.

The exciting implications of finding life elsewhere are what have fuelled the growth of the new discipline of astrobiology, described by one wag as 'the only science without a subject matter'. Astrobiologists are working on the assumption that if there is life out there we will be able to identify it as such, and it is astrobiologists who perhaps have the most vested interests in coming up with a workable and firm definition of what life is.

The suggestion that life is something that we will recognize when we see it is perhaps not popular with these people. In December 2006, Robert Hazen, professor of Earth Science at George Mason University in Virginia, wrote in *New Scientist* that 'I think the chances are good we *won't* [my italics] know alien life when we see it'.

A meeting held in the US that year to discuss the definition of life came up with answers that ranged from the sublime to the ridiculous. One expert on lipid molecules argued, rather parochially, that we must be searching for semi-permeable lipid membranes (which envelop every cell in the kind of life we know). Another expert on metabolism maintained that life began with the first self-sustaining metabolic cycle.

Yet others maintained that some sort of RNA coding was required, while a geologist sided with my old teacher and stated that life is essentially no more than a very elaborate self-replicating crystal. Some other definitions: 'Any population of entities which has the properties of multiplication, heredity and variation' (John Maynard Smith, the evolutionary biologist); 'an expected, collectively self-organized property of catalytic polymers' (information theorist Stuart Kauffman). Vaguer definitions include 'the ability to communicate' or 'a flow of energy, matter and communication'.

Robert Hazen points out that our total inability to define life is wholly unsurprising. After all, we have severe problems separating the living from the non-living, even when it comes to our own lives.

The issue of when human life begins in the womb is hugely contentious and has fuelled endless ethical and religious debates. A newly fertilized egg is clearly not the same thing as a baby, but equally clearly any difference between a 35-week-old foetus and a newborn infant is purely arbitrary. Finally, most clearly of all, assigning a definite date for the inception of life risks taking us into 'magic spark' territory.

'At the other end of the human journey', Hazen says, 'doctors and lawyers require a definition of life in order to deal ethically with patients who are brain dead or otherwise terminally unresponsive'. A century ago, you were dead when certain vital signs had ceased: when the heart stopped beating and when you stopped breathing. Now, in the era of heart–lung machines, intensive care and the occasional remarkable recovery from the deepest of comas, death seems to be more of a process than an event. The best definition we have seems to be 'you are dead when the doctors can do no more for you'.

In the end, a definition of life that distinguishes all imaginable living objects from the non-living ones is extremely elusive. Reliance on the existence of cell membranes, self-replication and so on seem to be arbitrary simply because we can all imagine things which we might all agree are alive and yet which do not fit these definitions. Gerald Joyce of the Scripps Research Institute has suggested a 'working definition' as simply 'any chemical system capable of undergoing Darwinian evolution'.

This definition, which is liked by Hazen, seems to include everything that we think of as alive, but would exclude things like computer simulations and machine consciousness, as well as artificial organisms designed to be incapable of adaptive evolution.

Maybe there is, in fact, no clear dividing line between 'clever crystals' and what we would call living organisms. Robert

Hazen says his favourite idea is that life on Earth may have arisen as a thin molecular coating on rock surfaces, a new form of complex mineral able to spread and grow a little like lichen (although this was most definitely *not* lichen) according to the nutrients available.

Having a decent definition of life will be necessary if we are to stand any chance of finding it elsewhere. Life could of course make its presence felt in a very obvious way. But assuming that little green men, ray-gun-wielding Martians and logical humanoids with pointy ears are, in galactic terms, thin on the ground, then we are probably talking about stuff that is a lot more subtle.

It may turn out, as some suspect, that life is fundamental. Life may be very, very old indeed, its precursors forming not long after the beginning of time, long before there were any planets on which it could get a grip. We may not live in a *Star Trek* galaxy, home to thousands of alien races with funny foreheads and a rather stilted grasp of English, but we may well live in a 'slime mould' galaxy where life, usually of the simplest of kinds, is staggeringly common.

Richard Taylor, the Secretary of the British Interplanetary Society, thinks that life may well be so common that there are at least ten bodies in our Solar System alone which have it. And if he is right about this – he is certainly not alone in his opinion – with the discovery that probably half of all stars in our galaxies have some kind of planetary accompaniment (this is a statistical likelihood basd on the number of extrasolar planets discovered in the last dozen years or so), the number of habitable – and inhabited – worlds in the Milky Way alone will run into the tens of billions.

Twenty years ago you would have found few scientists willing to speculate that there could be life elsewhere in our Solar System. This, before the post-1995 wave of discovery of extrasolar planets, made the hypothesis that we were probably very much alone a popular one. Now it is a quite commonly held view that there may be several 'local' candidates for life, and not just Mars, which has long been held to be the most likely second oasis in the Sun's family. At least one moon of Jupiter, a couple of Saturn's and maybe a whole new class of objects far beyond icy Pluto may plausibly be home to at least some sort of microbial life.

It is possible that precursors to life, like Hazen's rock-coatings, are common in environments where the suitable chemistry is available. Not the surface of the Moon, perhaps, but maybe on Titan, Saturn's now not-so-enigmatic largest satellite, where occasional cometary impacts and perhaps rhythmical climate change allow brief warming periods where such not-quite-life-and-not-quite-as-we-know-it chemistry could get going.

Finding fossilized or extant evidence for the precursors of life somewhere like Titan would be almost as interesting as finding life itself, for it would set a new boundary – the limits perhaps of the non-living. Titan has, on its surface, all the 'ingredients' (in warm-little-pond thinking) for life – lots of nice gloopy carbon compounds and water (albeit frozen) – and it is geologically active, with materials circulated from below the crust, to the surface, into the atmosphere and down again as precipitation, just like on Earth. Furthermore, although the surface of Titan is rather chilly, underground it may be quite different.

There is just a chance that in our lifetime some sort of conclusive evidence will be found for life elsewhere in our Solar System. The most likely place we will find it is on Mars, although it is just possible that something will turn up either on Europa, a moon of Jupiter thought to harbour an ocean of brine beneath its frozen surface, or Enceladus, another satellite of Saturn which we know has liquid water very close to the surface (the spaceprobe Cassini has photographed spectacular geysers erupting from this tiny moon, which is about the size of France).

Mars remains the most likely place that we will find evidence of extraterrestrial life in the near future. Of course no one is expecting little green men or the canal builders of legend, but recent discoveries from the Red Planet suggest that it may be a more amenable place, even today, for life than was thought even when the first space probes arrived in the 1960s.

In the late 1990s, a space probe called Mars Global Surveyor (MGS), now sadly defunct, sent back images of crater walls showing apparent 'gullies' and 'alcoves' a few hundred metres long and a few tens of metres wide, which the space-craft's imaging team said could only have been carved by running water. Over the years more gullies were found, together with thousands of strange dark streaks, all over Mars, that some said could only be caused by running water in the very recent past.

Then, in December 2006, an article was published in *Science*[1] showing comparisons between crater walls a few years apart using photographs taken by MGS. The pictures showed new gullies sprouting from the cliffs, the suggestion being that whatever is causing these features is causing them right now. Cue 'Water flows on Mars' headlines and even 'Has NASA found life on Mars?' in one newspaper, holding to the journalistic maxim that whenever a newspaper headline asks a question, the answer will invariably turn out to be negative.

Now, it must be said here that a few gullies do not mean that Mars is inhabited, even by hardy microbes. Perhaps too many scientists and journalists have rushed to populate Mars with life on the basis of what remains still rather flimsy evidence. I am sympathetic to the view of Australian geologist Nick Hoffman, whose 'White Mars' hypothesis states that many of the features we see on the Martian surface were carved by carbon dioxide flows rather than water, either now or in the distant past. 'We are in danger of looking at this planet through blue-tinted spectacles' he says.

We have, at best, circumstantial evidence that Mars has, or once had, conditions at or near its surface that would, at a push, be capable of supporting life for some of the hardiest micro-organisms currently living on Earth. But despite their sterling work, it is very unlikely that any of the current generation of robot space probes in orbit or on the surface of Mars will discover concrete evidence for life. To do this we will probably need to get hold of some Martian material and study it here on Earth or *in situ*, under the gaze of a human eye.

Actually, we may not need to go to Mars at all to do this. Mars (and the Earth) are being constantly bombarded by meteorites, some very large, capable of excavating impact craters and sending large amounts of Martian or Terrestrial material into space at escape velocities.

A decade ago there was a huge brouhaha when a meteorite (given the snappy label ALH 84001), found in the Allen Hills in Antarctica, was announced not only to have come from Mars but to contain curious tubular structures that were interpreted by some as being fossil bacteria. Since then, the verdict has swayed back and forth on ALH 84001; the balance of opinion now seems to be that the structures are not necessarily indicative of life, but no doubt the pendulum will swing back once again.

What does all this tell us? Well, firstly, if we can show that life is present, or has been present (the structures in ALH 84001, whatever they are, were 'fossilized' several hundreds of millions of years ago) on at least one other body in our Solar System we will have to draw one of two conclusions. One is interesting, but not a paradigm shift. The other is both. And the thing is, we may never know which conclusion to draw.

The first conclusion would be that life arose somewhere in the Solar System and migrated, using the interplanetary meteor courier system, to other bodies. Maybe, as the Australia-based physicist Paul Davies has speculated, the original genesis could have happened on Mars. Only later did life migrate to Earth.

There is circumstantial evidence that this could be the case. Mars, being smaller, would have suffered less of the hellish bombardment that scarred the planets and moons of the Solar System in its early days (Mars had less gravity and was, simply, a smaller target). Life would have stood a better chance of getting a foothold and surviving there early on than on the (relatively massive) Earth. It is possible also, that with a thinner atmosphere and less Solar radiation, the early Martian environment of four billion years ago was more hospitable than that of the early Earth. If this is the case, then we would all be, as Davies points out, Martians. The War of the Worlds would have been won four aeons ago.

Conversely, life could have begun on Earth and spread to Mars. Or it could have begun on Earth or Mars, and spread from there to the outer Solar System. If we find signs of life on, or under, Europa or Enceladus or Titan, we could be conceivably looking at Martians or Terrestrials. It is also possible (although unlikely perhaps, given the distances involved and the frigid temperatures pertaining out there) that life first arose on, say, Titan, and migrated to Earth. We will only know if we manage to get hold of one of these aliens and do a DNA analysis. If we find extant life on Mars and discover that it is related to life on Earth, it would be a fascinating finding, but it might not actually say much about what life is, how it arose and how common it is in the Universe. We would simply be living in a Solar System full of our cousins.

But say we found life on Mars and, furthermore, discovered that not only was its genetic material completely unrelated to that on Earth, but that Martian life didn't even use DNA at all, but some completely different chemical. Then we would be forced to conclude that life had arisen, independently, at least twice in our Solar System.

We could then conclude that since the first place we went looking for it, the nearest likely habitat for life, has undergone independent biogenesis, and furthermore since this planet is quite different in fundamental ways from our own, life is probably not only common but ubiquitous in the Universe.

But there remains a third possibility. That life arose on Earth not as the result of some sort of cross-contamination with another nearby planet (or vice versa) but that it arrived in the Solar System from somewhere else entirely.

The idea of *panspermia* was first mooted by the Greek philosopher Anaxagoras, but although long dismissed as unlikely, it has never quite fallen to the pit of true scientific disreputability. It states, simply, that life is scattered throughout space as 'seeds' or 'spores' which propagate throughout the cosmos.

A weak version of the hypothesis, interplanetary *lithopanspermia* envisages the situation outlined above – that life could arise, once or more than once, on one planet in our Solar System and spread to other bodies in the Solar empire carried on space rocks (or even space probes in which humankind may have unwittingly sparked panspermia in our own neighbourhood very recently). A stronger interstellar version posits that life could spread between star systems using essentially the same means of transport, bits of rocks blasted off planetary surfaces during meteor or cometary impacts.

A more profound kind of panspermia holds that life, or its antecedents, permeates space, perhaps in cosmic dust, or tied up in the swarms of icy bodies that probably envelop every star and are scattered in the voids between. The strongest version of all posits that life is actually a fundamental property of the Universe, and owes its origins to processes and events that arose during or shortly after the Big Bang itself. According to this hypothesis, 'life' is every much as fundamental as, say, the strong nuclear force or the gravitational constant.

Before deciding which, if any, of these possibilities is most likely to be correct, consider the implications. Firstly, the panspermia hypothesis does not really answer any questions about how and when life began. It merely pushes the date of biogenesis back to before the beginning of life on Earth. It also does not speculate on whether life arose just once in the early Universe or many times at separate locations.

One variant of the panspermia hypothesis states that out there somewhere, intelligent life is deliberately seeding the cosmos by firing huge quantities of DNA into space. This again does not solve the genesis problem, and sounds pretty outlandish, but it has been taken seriously by some serious scientists, notably Francis Crick, the co-discoverer of the helical structure of DNA. Panspermia seems, at first glance, to be an unnecessary elaboration on an already elaborate problem.

But there are several strong arguments for panspermia which are hard to totally refute. Perhaps the most convincing 'evidence' is the extremely short – some say suspiciously short – time it took life to arise on Earth after it was formed. Greenlandic rocks of great antiquity, some 3,850 million years old, have been found containing bounded iron formations thought to have been released by photosynthetic plants.

Less controversially (these iron deposits could conceivably have a non-biological origin), *stromatolites* – fossilized marine bacterial colonies (living examples can be seen today in Western Australia) have been found that are 3.5 billion years old. The point is that the Earth is around 4.55 billion years old, and it is believed that for the first few hundred million years or so of its existence it suffered a bombardment by Solar System debris on a scale millions of times more frequent and violent than we see today.

Every few tens of millions of years the Earth would have been hit by a rock large enough to effectively sterilize it. One impact, by a Mars-sized object, is thought to have blasted a huge molten ball of magma into space which cooled and became the Moon.

The vital thing is that all life on Earth today must be descended from an organism that came into being *after* the last sterilizing event (it is quite possible that biogenesis in fact occurred several times, each lineage being wiped out by impactors in this period, appropriately named the Hadean, in which case our oldest fossils may effectively be those of alien life forms). This doesn't give very much time for life to get going, a few hundred or even a few tens of millions of years at most. To some, this is implausible. The Universe as a whole has been 'life friendly', in terms of the requisite chemicals being available, for a much longer time. It is perhaps statistically more likely, say the panspermians, that the 'original' life arose during this much longer time period than during the rather limited window afforded on Earth itself.

There is another piece of evidence that counts both for *and* against panspermia, and that is the discovery that life on Earth can thrive in a far greater variety of environments than was once thought possible. Again, at school, we were taught that life was pretty much impossible at temperatures over about 60 °C. Hotter than this, and proteins denature and DNA falls apart. Now we know that some species of bacteria not only survive but thrive in temperatures hotter than the boiling point of water, clustered around the deep ocean vents known as black smokers. Everywhere we look, we find life: deep under the Antarctic ice and, most importantly perhaps, several kilometres underground.

There is some evidence (although this is as yet not widely accepted) that tiny organisms, putatively named nanobes, may be able to survive 10–20 kilometres underground under conditions of really quite extreme heat and pressure. Perhaps most pertinently, viable spores have been discovered in mineral grains that are hundreds of millions of years old. Life can not only tolerate high temperatures but extreme pH levels too, as witnessed by the discovery of micro-organisms which happily thrive in hot springs of basically sulphuric acid.

And life can cope with very cold conditions too. Deep in the polar ice sheets a bacterium called *Colwellia* has been found at temperatures of –40 °C, and in the lab these organisms have been found capable of surviving the temperatures of liquid nitrogen (–196 °C), a finding which Dirk Schulze-Makuch, an astrobiologist at Washington State University described to *New Scientist* in early 2007 as 'mind boggling'. Until recently the lower limit for life was thought to be around –30 °C, and the discovery of *psychrophiles*, cold-loving organisms, has come as a real surprise. For a start, it demonstrates the impressive and elaborate tricks that such organisms have up their sleeves to prevent the water in their cells from freezing. By concentrating salt, the freezing point of water drops to –50 °C. Sticky proteins called exopolymers can prevent the formation of cell-shattering ice crystals. But just how *Colwellia* can not only survive being immersed in liquid nitrogen but actually appear to metabolize at these temperatures remains a mystery.

The fact that life can survive at such low temperatures does, if anything, extend the possible range for life in the Universe generally far more than the discovery that some like it hot. After all, much of the Universe is extremely cold, not boiling hot. Some of the more promising homes for life in our Solar System are the large satellites of the outer planets where it is very frigid indeed.

If microbes are happy at –40 °C then suddenly whole swathes of Solar real estate, including the vast, uncountable swarm of icy objects out in the Kuiper Belt, beyond Pluto, suddenly look far more hospitable (it is thought that the interiors of these objects may be kept warm by the heat of radioactive decay, even if their surfaces are 200 degrees below zero or colder). Forget Richard Taylor's 10 possible homes for life; the true number of places where microbes could eke out a living in our Solar System alone may run into the hundreds or thousands.

The discovery that life is more robust and adaptable is, at one reading, evidence against panspermia, as it suggests that we may need to examine a far greater range of possibilities for the location of biogenesis on Earth than we thought. It may well be the case, for instance, that we will have to rethink the assumption that during the first few hundred million years of its existence the Earth was uninhabitable; if bacteria can survive 20 kilometres underground it is possible that life could have clung on even during the most severe batterings of the early Hadean. The existence of the extremophiles also muddies the water considerably when considering what is to be thought the 'normal' habitat for life on Earth; is it on the surface, in the oceans and in the air as we (and Darwin) thought, or is there a hugely bigger hidden biosphere underground, in the rocks and buried deep in the sediments, under the ice or in the upper reaches of the atmosphere? Did life evolve on the surface and migrate downwards, or did it evolve underground and move up when the skies cleared? Did life possibly evolve in the sky? We may never know.

But while extremophiles seem to suggest that it may have been easier for life to get a hold on earth in its earliest days, they also point to the possibility that life – microbial life and spores – is robust enough to survive the rigours of interplanetary and even interstellar space, as the panspermia hypothesis suggests. The cold of space holds no barrier when a microbe which evolved on the balmy Earth can continue functioning at −200 °C.

This is all circumstantial evidence for panspermia, however (just as it is circumstantial evidence to posit that because some microbes *could* live on Mars then microbial life will by dint have evolved there). Do we have any reason to believe that life actually *has* arrived on Earth?

The champions of modern panspermia were the late Sir Fred Hoyle and his colleague Chandra Wickramasinghe, now at Cardiff University. As well as being Big Bang sceptics, Hoyle and Wickramasinghe held that not only had Earth been seeded by life from space billions of years ago, but that active spores continue to rain down upon the planet today. This could explain, they said, the often mysterious epidemics that plague humanity. Around 40,000 tonnes of carbonaceous material fall onto the Earth each year from space, and Hoyle maintained that about a tonne of this was in the form of actual bacteria or bacterial spores. In 2003, during the height of the SARS epidemic in Asia, which killed several hundred people, Wickramasinghe wrote to *The Lancet* medical journal claiming that the virus responsible was possibly extraterrestrial in origin.

This belief – and it *is* a belief, and a fringe one at that – stems from the discovery that complex organic (carbon-containing) compounds are common and exist in large quantities and great variety in a number of (on the face of it) rather unpromising cosmic objects, most notably comets. Astronomers have detected, using spectroscopy, many kinds of organic molecules in space, floating in clouds of gas or bound up in dust particles. They range from simple compounds like methane, hydrogen cyanide and alcohols, including ethyl alcohol, to more complex molecules, such as amino acids, of which more than 70 have been found in meteorites.

An experiment performed by NASA in 2001 attempted to replicate the effect of a comet, loaded with amino acids (the building blocks of proteins) slamming into Earth at thousands of kilometres an hour. Rather than fragmenting the amino acids, as was assumed would happen, the impact (using a sort of high-velocity bullet) turned out to actually force the amino acids to link together to form peptide chains, polymer compounds just one stage less complex than proteins themselves.

In fact, a class of compound called nitrogenated aromatics has been found just about everywhere we look in space: in comets, in interstellar dust clouds, and in the atmospheres of the outer planets. These compounds – carbon-based molecules based on a ring structure – are, generically, the building blocks for life, forming the basis for complex biological compounds such as proteins and nucleic acids.

In February 2004, Professor Sandra Pizzarello and colleagues published a paper in *Science*[2] in which she argued that the *chirality* – the tendency for the molecules to be left- or right-handed – of proteins and sugars in Earth life could be linked to the meteoritic material which has hit our planet over billions of years.

A class of meteorite called carbonaceous chondrites contains the most complex carbon compounds known from outside Earth, including amino acids (the building blocks of proteins) and sugars. Pizzarello found that in experiments where sugar synthesis was performed in the laboratory under conditions thought to have pertained on the early Earth, a constant rain of chemicals of the right (or in this case, literally, left) 'handedness' caused the 'native' sugars to change their chirality. This does not mean that Pizarello has proved that life arrived on meteorites, but that it is at least possible that the arrival of meteorites might have had a profound influence on the evolution of that life, however it got started on our planet.

We know that these organic molecules have been present in the Universe for a very long time, certainly predating the existence of the Earth. A prime candidate for the 'oldest thing on

Earth' is a meteorite which slammed rather dramatically and very visibly into Lake Tagish in Canada in 2000. In 2006, details of the analysis of the Tagish meteorite were released in *Science*[3], revealing that the grains from which it is composed predate even the formation of the Sun. Hollow carbon spheres found in the meteorite fragments, each a few thousands of a millimetre across, were speculatively dated as being several billions of years older even than the 4.6 billion years of our star. It is possible, in other words, that contained in this rock fragment are particles almost as old as the Universe itself, containing complex compounds, including amino acids, interleaved with clay mineral grains.

The clay minerals, silicates arranged in layers, are (it is speculated) possible 'wombs' in the formation of some sort of pre-life entity, maybe complex self-replicating proteins or the precursors to nucleic acids. Commenting on the finding, one of the investigators, said: 'These things tell us what kinds of chemicals are out there in interstellar space. They could have been the original seeds for life to get started'.

The idea that life, or at least the chemical precursors to life, could have arrived on the Earth carried by cometary material is no longer regarded as outlandish. Comets are deeply mysterious, part of a class of ubiquitous icy objects that may envelop every star system and form on the edges of interstellar clouds. Comets and their relations in the Kuiper Belt may be ambassadors from a time long before our Solar System was formed. The discovery of complex organic chemicals on meteoritic fragments strongly suggests that at least some of the processes once assumed to have taken place in Darwin's warm little pond may instead have occurred in deep space, at a time when the Universe was still young.

This is what we know; from now on all is speculation. As far as 'evidence' for more complex life forms on distant planets, discussions of UFOs and their inhabitants belong somewhere else. But it is possible to speculate nonetheless on scenarios that are just as weird and wonderful as flying saucers. If life really owes its origins to a time in the very early history of the Universe, does this suggest that life is possibly an intrinsic property of our cosmos?

On the wilder shores of the panspermia community it has been suggested that life may itself form part of the 'organizing principle' of the Universe, with its origins in the Big Bang itself. This can be seen as an extension of the 'Gaia' concept, advocated by the British biologist James Lovelock, which states that living and non-living processes on Earth are locked into a series of feedback mechanisms. In other words, the Earth is like it is because of the life that is on it, as much as the other way round.

In an article in *Nature*[4], in 2004, William Dietrich, of the University of California at Berkeley, speculates what the Earth would be like if it had never had life. 'Very little has been written about the evolution of Earth in the absence of life', he writes. He goes on to speculate that life on Earth may have had far more profound effects that simply altering the composition of the atmosphere (pumping in oxygen via photosynthesis) and the formation of fossiliferous rocks:

> On Earth, plate tectonics depends on an upper mantle's low-viscosity zone on which plates can glide, and it has been proposed that this zone arises from the injections of water at subduction zones.... Is it possible that the emergence of life on earth prevented the development of atmospheric conditions favourable to solar wind erosion, keeping the planet 'wet' and enabling plate tectonics? Is plate tectonics on Earth a consequence of life on Earth?

If this hypothesis is right, it would be an extraordinary thing to say about our planet. Life may have saved the Earth from turning into another Venus: by squirrelling away gigatonnes of carbon dioxide in their shells, countless quadrillions of tiny sea creatures could, over the aeons, have prevented a massive CO_2 build up in Earth's atmosphere and a consequent runaway greenhouse effect, which seems to have been what happened on Venus. Earth could then really be considered, without any of the perhaps unfortunate spiritual overtones generated by the original Gaia concept, a 'living' planet, a biological system, with the same relationship to the organisms that live on and in it as a shell to the snail within.

Could the Universe be like this, its properties at least partly influenced by the presence of life? Certainly, on a very fundamental level the anthropic principle states that the Universe is a 'Goldilocks universe', fine-tuned to be just right for life. The physical constants, the nature of matter and the particles and so on seem to be set up in such a way that they allow life to be viable, and – the important bit – it seems there is a very narrow 'window' outside which we would certainly be looking at a lifeless universe.

The physicist Lee Smolin proposed a theory back in the 1990s that one possible origin of the Universe – that it arose as a bubble of space–time in another universe created by a black hole – suggests that life-friendly universes may be subject to a form of cosmic natural selection; in short, universes which are capable of producing the maximum number of black holes would be the most productive 'parents', and it is the case that a universe set up to produce black holes (lots of stable stars of the right size and mass) are also universes where the physical constants permit 'our' sort of life to exist.

This theory suggests that life, if not an intrinsic property of the Universe, is at least central to our understanding of why the world around us is as it is. The exciting thing is that so much of this is testable. We can search for evidence of life on other planets and in the debris and detritus that hits the Earth.

If the Universe does turn out to be alive then it would, in philosophical terms, not only be an outstanding discovery but it would also turn the heat off humanity a bit. We may never know how frequent intelligent life is, but at least we would know that should we inadvertently bring about our own demise there would be an awful lot of places out there where the great life project would carry on. That warm little pond may have been a very big sea indeed.

References

1 Malin, M. C., Edgett, K. S., Posiolova, L. V., McColley, S. M. and Noe Dobrea, E. Z. (2006) Present-day impact cratering rate and contemporary gully activity on Mars. *Science*, **314**, 1573–7.

2 Pizzarello, S. and Weber, A. L. (2004) Prebiotic amino acids as asymmetric catalysts. *Science*, **303**, 1151.

3 Nakamura-Messenger, K., Messenger, S., Keller, L. P., Clemett, S. J. and Zolensky, M. E. (2006) Organic globules in the Tagish Lake meteorite: remnants of the protosolar disk. *Science*, **314**, 5804.

4 Dietrich, W. E. and Perron, J. T. (2006) The search for a topographic signature of life. *Nature*, **439**, 411–17.

7

are you the same person you were a minute ago?

What a silly question. Of course you are. You have the same skin, the same bones, the same meat. More importantly, because of course we are talking about your mind here, the essence of your good self, you have the same brain: the same (more or less) neurons, the same synapses, the same blood vessels and connective tissue inside your skull. More profoundly, you have the same memories. And unless something alarming happens to you, like an almighty thwack on the head, you have the impression of an unbroken continuum of consciousness.

But the more you think about it, the more you realize that you are not the same person you were even a second ago. You may be composed of the same atoms, but only approximately. In one minute, for example, you will metabolize about 4.5 grams of oxygen – that is one quadrillion atoms or thereabouts every 60 seconds. Several trillions of carbon dioxide molecules leave your lungs and trillions more will be assimilated from the products of digestion. In short, the cellular chemistry that makes you tick is a constant merry-go-round of breakdown and synthesis. On a molecular and atomic level you are not *quite* the same person you were a minute ago.

Forty years ago I laid claim to about 10% of the mass that I am now and more or less every atom in my body has been replaced since then. In another 40 years' time, if I make it that far, I will be similarly different from today. And it is not just the material structure of my body that changes, every year, every hour, every second. I am moving, constantly and unstoppably. If I sit as still as I can I am still rotating at about 450 mph around the Earth's axis, and at 66,000 mph around that of the Sun. Every microsecond I occupy another wodge of space–time. My makeup and position is changing all the time. If 'I' have any definition it is that I am an indeterminate bag of flesh and bones that walks around claiming to be me.

The concept of self, and the continuity of existence, is an old philosophical chestnut. At its root is the thorny issue of consciousness, the mysterious feeling of self-awareness that remains totally unexplained. The practicalities of identity are something that we think about more than ever before. This is an age when governments, in the dread 'interest of security' want to fingerprint us, photograph our irises, and barcode our 'biometric' data and add this to a database containing our every movement, caught on CCTV and trawls through cyberspace.

In a banal way, our identities are thus reduced to a stream of numbers. With the advent of high-speed Internet access, millions of people are establishing second, third and more identities in the online world, something that has been endlessly predicted since the advent of the Internet 30 years ago and yet which is only now starting to come true.

Millions of people now have 'lives' in electronic worlds like Second Life, and spend considerable amounts of time, effort and money pretending to be someone else and living an artificial life. It is easy to see this trend continuing and these alter-egos proliferating.

Back to reality. Establishing a concept of identity matters in so many ways. Courts, as never before, recognize the truth that the person before them, while sharing the same body, may not be the same person who committed the crime. Madness, physical illness and profound injury may intercede to create a new person. That much has been recognized for a long time. But the more science gains an understanding of what the mind is (which admittedly isn't much of an understanding as yet), the more we are forced to consider the possibility that the continuity of existence is a fiction.

Nowadays, a schizophrenic who kills is excused prison on account of his condition. How long before a man, caught for a crime he committed 40 years ago in his teens, might reasonably argue that he is simply not the same person that he was then and thus also escape punishment? Pinning down what we mean by identity has profound implications for the way we treat criminals or the elderly, and for how we think about ourselves. What light can science shine on the matter? And does establishing what we mean by identity and self have any relevance to the 'hard' consciousness question?

There are two basic views on identity. One, the 'folk science' view, holds that there is an 'inner essence' of personhood, an 'ego' that is constant throughout time. This is the way most of us think about identity. I am the same person I was as a child and I shall continue to be the same person until the day I die. Yes, my body changes but some sort of 'essence of me' remains constant. The problem, comforting and coherent as the ego theory is, is that it cannot be true. Not literally.

The alternative idea, that we are a 'bundle' of mental states, tumbling down time's motorway like a tumbleweed blowing in the wind, offends just about everything we think to be true about ourselves. We are not what we believe ourselves to be. Our lives are a series of related experiences, but there is no single entity at the centre actually having these experiences. This is a disturbing, even distressing, viewpoint. It says that there is no real self. That continuity of existence is an illusion (although the consciousness of those experiences is not). Before we can consider just what it is that is generating this experience of endless 'nows', we need to put the old ego theory to rest and kill the ghost in the machine.

The problem with a 'soul' is that we need to describe what it is, how it arises and how it interacts with ordinary matter, i.e. the cells of our brain. There is no way of doing this and no evidence for the existence of such a phenomenon. Science now rejects the idea of the soul, on the basis of Occam's razor as much as anything. Rather than believing that a soul (for which we have no evidence) uses the brain for thinking, it is far simpler to believe that it is the brain itself that is generating the experience of those thoughts.

A telling illumination on what the folk science concept of self actually means was made by the philosopher Derek Parfit. In his 1986 book *Reasons and Persons* he dissects the commonly held notion of the self with a series of neat thought experiments involving teleport machines. Teleporters, like time machines, are useful devices, not just as fun sci-fi plot devices, but because even before anyone has built one and switched it on they can be used to perform all sorts of interesting mind experiments.

The classical sci-fi teleport machine works in one of two ways. In the first, the object to be teleported, say a human being, is somehow transported *in toto* through the ether to her destination. These machines make use of exotic structures like wormholes, able to warp and bend space to allow instantaneous travel from point A to point B. While interesting, these teleport devices do not pose interesting philosophical questions. They are, essentially, trucks, albeit very exotic and clever ones, because the essence of what is being transported is not interfered with in any way during the process, any more than if you were to take a trip by aeroplane.

The interesting questions arise with the other sort of teleport machine, the sort that takes you to pieces and then reassembles pieces exactly like those original ones somewhere else. No one knows whether it is possible to construct such a machine (in fact there is some experimental evidence that building a teleporter should be considerably easier than a time machine), but the feasibility of such a device is not important to our thought experiment.

In one scenario, you step into the teleport booth and your body is scanned with exquisite precision. Each of your 7,000 trillion trillion atoms needs to be plotted, by type and position. It is not clear how this could be achieved. Possibly some sort of refinement of the X-ray crystallography technique. The machine also needs to know the precise quantum state of every particle. Recent experiments, where whole atoms have been successfully teleported across a room, suggests that should a teleport machine capable of handling larger objects such as a person be possible, it will probably be impossible to complete this scanning (and subsequent transport) without destroying the 'original'. Hopefully, the scanning process would be near instantaneous and painless.

Then the information gleaned by the scanning machine must be transmitted to the destination. Again, experiments suggest that this may be possible using one of the stranger beasts in the quantum menagerie, the phenomenon of entanglement. This allows the properties of one particle to be transmitted an arbitrary distance instantaneously. This seems to break all sorts of laws, not least the commandment which says that *thou shalt not travel faster than light*, but physicists insist that they have all the bases covered. It can be done, for entire atoms at least, as proved in 2004 by Professor Rainer Blatt in Austria and his colleague Dr David Wineland in the US, who managed to teleport entire atoms across the lab.

There are several possible scenarios explored by Parfit and all have disquieting implications. If the teleport machine works as it should, a precise copy, complete with your memories, walks out of the receiving booth. The original is destroyed. Most people would be happy to say that the machine has 'worked' – that the original 'you' has indeed been moved from A to B.

But it is when the machine is tweaked, or malfunctions, that it becomes clear that something quite profound is being illustrated. Say the machine makes two copies. Which one is 'you'? Say the original 'you' is not destroyed. Who can lay claim to the ownership of your 'essence'? Once you start thinking about this you conclude, as Parfit does, that teleportation is murder. These machines are simply making replicas. But then you can also go on to conclude, as Parfit does, that this doesn't matter. The new you is not an impostor even though the old you is dead. There is no paradox here because the old you is being destroyed, in your own brain, all the time anyway.

Enough of teleport machines and back to the real world. Say, for example, you have a terrible road accident. In the accident you receive a bad head injury, so bad that for a few minutes your brain function drops to nearly zero. But luck is on your side. The inflammation subsides and the surgeon is good. Damage is minimized. Nevertheless you spend several weeks in a profound coma. You have no waking thoughts. You do not dream. Magnetic resonance imaging shows your brain to be just meat. Live meat, but not thinking meat.

Happily, your body rallies round. Blood vessels regrow, synapses start to fire. A month or two later you wake up. For some time you are not yourself. Your memories are patchy. You have trouble remembering some facts about your life, but your memories return. Helped by friends and family you painstakingly reconstruct your past, filling in the gaps in the narrative of your life. Within a year you are well enough to return to work; within two, save for some nasty scars, you are effectively back to normal.

Or imagine another accident. In this case, a real one, which happened to a real man, one Phineas Gage. He was working on a railway construction site as a foreman in 1848. He was part of a team setting explosives to blast a way through a rocky outcrop. Something went terribly wrong and a piece of metal tool a metre long and weighing more than six kilos was driven by the dynamite through his skull and frontal lobes, the key bits of grey matter at the front of his brain.

Everyone assumed he was dead, but by some strange happenstance he was not. Better still, he made a more or less full recovery. But something was not the same. Accounts vary, and recently some doubt has been thrown on this story, but there are claims that Gage had a fairly dramatic change in his personality. His intelligence and his memories were largely intact. Yet according to his doctor and family, the formerly hard-working, mild-mannered and temperate man became prone to bouts of drunkenness and bad temper.

Finally, something altogether more mundane: sleep. You drift off at the end of a long day – a perfectly natural process that involves neither teleport machines, horrible road accidents nor metre-long metal spikes. Within minutes you have entered another world, the world of the sandman. Sometimes you dream, sometimes you do not, but you are not conscious, at least not in the way that you are during the day. You have not, unless you are very lucky, any wilful control over either the subject or direction of your dreams. Your brain, while not in the profoundly altered state that is the coma, is nevertheless not 'itself' while you are asleep. 'You' effectively switched yourself off for a few hours, for purposes that are still not entirely clear. When you awake you of course have all your memories, atoms and so on.

These examples throw up a profound riddle about the nature of our existence. What does it mean to be 'me'? And what do I mean by the continuation of my identity?

The teleport machine has to destroy you to make a replica, but that replica has identical memories to the original. Imagine, in a final twist, that the teleport machine makes a copy, but in exactly the same place as you were when you stepped in. Furthermore, the destruction–recreation process takes an extremely short time – a few trillion-trillionths of a second – far shorter than the time it takes anything to happen in an intact brain. You have effectively been killed and rebuilt in less time than it takes a photon to cross the diameter of an atom. Would you notice?

Phineas Gage was a changed man, but he inhabited the same body. When you sleep you turn off part of your brain, which is then revived in the morning. What is the difference between this and teleportation? When you suffer from Alzheimer's, you lose memories. Are you losing the essence of self? And most people have forgotten anything that happened to them in the first few months of their existence. Does this mean that they were a different person then?

To see how slippery identity can be, consider the case of false memories. If there is an essence of identity it is our memories which come closest, but it is quite possible, inadvertently or deliberately, for people to acquire memories of events which never happened.

Several studies have shown how easy it is to create false memories. In fact, many of our memories are 'false' in that they do not accurately represent real events which could be corroborated by witnesses. Most people have memories of childhood happinesses and traumas that may be confabulations generated by real memories and later descriptions by our parents.

Just how easy it is to implant false memories has been shown by several scientists, most famously the American psychologist Elizabeth Loftus. In the 1974 'Reconstruction of Automobile Destruction Study' Loftus showed that an eyewitnesses memory can be easily altered by information supplied to them after the event.

Student volunteers were shown film of car accidents and asked to write reports on what they had seen. Specifically, the students were asked to gauge the cars' speed after being asked the question 'About how fast were the cars going when they [verb] with each other?' The missing verb was either 'collided', 'smashed', bumped', 'hit' or 'contacted'. They found that the 'faster' the verb used to describe the collision the faster the reported speed. Collisions where the cars 'smashed' were estimated to have taken place nearly 8 mph faster than ones where the cars merely 'contacted'.

A second study was done, and this time after being shown the collision the students were asked to say whether or not there was broken glass at the scene. The more 'rapid' the verb used by the experimenters to describe the collision, the more likely the students were to 'remember' seeing smashed glass (in fact there was none). When the verb 'smashed' was used more than a third of the volunteers 'remembered' seeing glass that wasn't there.

The importance of these findings is clear. Most obviously there are ramifications when it comes to court evidence and statements given by witnesses to police officers. It is now accepted that being an eye-witness to an event is no guarantee whatsoever of accuracy. It is also clear that the way any question is phrased, by a police officer or a barrister, say, can have a profound effect on what we 'remember'.

False memory syndrome (FMS) is an even more profound distortion of 'self' than simply misremembering a car accident. FMS describes a state of mind in which 'sufferers', if that is the word, have vivid but entirely inaccurate memories of childhood trauma, most notably sexual abuse. The accusation is made that some professionals, hypnotists and psychologists suspicious of child abuse unwittingly 'implant' such false memories in their patients, with disastrous consequences. The patients may now believe that they had been abused by a parent, resulting in families breaking up and people going to prison.

In the 1980s and 1990s, a series of cases made the headlines in North America and in Europe where large numbers of adults were accused of engaging in ritualistic or satanic sexual abuse of children in their care. Although child sex abuse is distressingly common, satanic abuse is not. Nevertheless, the numbers of people affected ran into the hundreds and a lot of people were incarcerated.

In the enquiries, court cases and lawsuits that followed, the extraordinary and disturbing lengths to which some social workers and psychologists would go to implant memories in young minds was revealed. The trauma of the accusations, which often led to the break-up of families and wrongful prosecutions has been well documented; what has been less well documented is what amounts to the destruction of the self for some of these children. Individuals with happy memories of their parents and loved ones have been replaced by unhappy individuals who 'remember' horrid events which did not happen – disturbances to the self which remained in place even after the falsehood of the claims was revealed. In some senses, new individuals have been created out of nothing.

Less serious is the strange phenomenon of alien abduction. It is hard to know just how many people now believe that they have been abducted by aliens, taken aboard their spacecraft and subjected to often humiliating medical experiments, but some estimates put the numbers well into the hundreds of thousands or more. One survey suggested that 1% of the US population (3 million people) are convinced that they have been abducted – an extraordinary statistic if true. This is not a uniquely American phenomenon, but Americans certainly dominate the lists of abductees.

Sometimes people have continuous, unbroken memories of experiences which include alien abduction. Far more often, the abduction event is 'revealed' through the medium of therapy or hypnosis. The experience is seen by many as key to understanding whatever neurotic conditions may be troubling them, and therapists assure them that the recollection of being 'taken' is cathartic and therapeutic.

Of course, we have no evidence that any human being has ever been abducted by aliens, so we must assume that these recollections are bogus – either these people or lying, or they have had memories – false fragments of self – implanted by their therapists.

What is really interesting is that the inauthenticity of these memories doesn't seem to matter. At the American Association for the Advancement of Science's annual meeting in Denver in 2003, Harvard psychologist Richard McNally presented findings that showed that people who believed they had been abducted by aliens suffer 'real' symptoms such as disturbed sleep patterns, 'underscoring the power of emotional belief'.

Many of the 'abductees' have a lot in common – such as a whole host of related 'fringe' beliefs: past lives, astral projection, tarot cards, the occult and so on. McNally calls this a 'common recipe'. Many also have episodes of sleep paralysis and hallucinations, which are often what prompted them to visit a therapist in the first place. It was these therapists who usually suggested alien abduction as an 'explanation' for their symptoms.

These recollections became real. War veterans who have suffered traumatic experiences on the battlefield show characteristic physiological reactions when shown footage of combat. Their heart rates go up, they sweat, and they breathe faster and more shallowly. People who have not experienced combat do not show the same responses. But 'alien abductees' show at least the same responses – increased sweating and heart rate and so on – as Vietnam vets when told tales of being abducted and having unpleasant experiments performed upon them aboard alien spacecraft.

The nature of identity, particularly the continuation of identity, is in fact a classic philosophical problem that has been debated at some length. In 'The Possibility of Altruism', the philosopher Thomas Nagel points out: 'The idea of a temporally persistent human being is an exceedingly complicated one'. It leads to other dilemmas. What weight should we give our future and past selves? Should we treat them as separate entities and indeed treat one better than the other? It seems strange to ask this question, but in fact most of us do make a distinction between them.

As Nagel says, we should have reason to regret our bad behaviour at the office party last night, not just because the unfortunate possible consequences now and in the future, but also because the bad behaviour was bad then *per se*.

The person who you were then has become the person you are now, and that regret is a link between the two. Similarly, we should take account of our future life in a similar way to how we would consider the life of a different person. By deciding to take up smoking, or allowing myself to become a drug addict, I may be said to be causing harm to a future 'me' that is not really the me of today. The first cigarette will not kill me, nor the first injection of heroin, but in 20 years' time the person who calls themselves me may have cause to regret the decision made by me now. Becoming addicted to nicotine is thus less like suicide than murder.

That is the philosophy, but what about the science? As we have seen, the idea of continuous unbroken identity makes little sense unless you invoke some sort of ghost in the machine, an old-fashioned soul. But we have no evidence that this is the case. Instead, there is no 'self', just a series of patterns of information which can be created and destroyed over and over again. Paul Broks, a psychologist at Plymouth University in the UK, wrote in *New Scientist* in 2006, 'There is no self to destroy. The patterns are all'.

This leads to what Broks calls a 'neat inversion' of conventional thinking. 'Those who believe in an essence, or soul, suddenly become materialists, dreading the loss of an original body [say in a teleport machine]. But those of us who do not hold such beliefs are prepared to countenance life after bodily death'.

Even without the trauma of hypnotically induced false memories, serious brain injuries or the altered states created by certain narcotics, all of us will have had at least two 'selves' in our lives. In February 2007, Patricia Bauer of Duke University presented findings to the AAAS annual conference in San Francisco that suggest that human infants have a totally separate 'self' from their later adult forms. The phenomenon of 'infant amnesia' is puzzling; why are the vast majority of people completely unable to recollect any events in the first year or so of their lives? One explanation is that infants are simply unable to lay down memories. But Bauer has shown, by attaching electrical sensors to the infants' scalps to record their responses to sounds and pictures, that infants do form memories in their first year in much the same way as adults. The infants in the study were able to recall events over periods of days or even months. But because of the immaturity of their brains these memories evaporate, uncommitted to long-term storage; the old self is lost and a new one is born.

The realization that the self is not immutable has practical as well as philosophical implications. Once we are free of the shackles of the soul, we are also free of having to think about the self as an inviolable entity. Seeing how easy it is for memories to be created (by suggestive therapists and lawyers) and destroyed (by illness or trauma) should make us sceptical of any account of any controversial event such as a crime, where accuracy is, or should be, all.

Not only are our selves dying and being reborn every second due to the normal processes in our brains, they are also being augmented, sculpted and remade by the people around us and by our own agency. Now that it is clear just how easy it is to implant traumatic memories in the minds of children, we see far fewer families broken up after unfounded and false claims of ritual abuse than was the case 20 years ago.

Our self is not only being broken and constantly remade, it is also now clear that it cannot be distinguished from the selves of others as clearly as was thought, meaning that it is really true that no man is an island. In the 1990s, it was discovered that certain brain cells, called 'mirror neurons', were present in the brains of primates (and probably also humans and birds). Found in the ventral premotor cortex, these cells fire when monkeys perform certain tasks and also when the animal watches another perform the same task. Functional MRI scans of human brains have shown similar systems in the human brain.

What does this mean? It means that the brain is constructing a model of the 'self' that is in some way outside of itself. Mirror cells allow one 'self' to create a 'bridge' to another. Mirror cells have been implicated in the development of language, and also in the development of social networks.

Mirror cells have also been implicated in the development of a theory of mind, knowing (or guessing) what another is thinking. It is possible that a deficiency in the mirror cell system could be responsible for autism, as autists seem in many cases to lack a theory of mind. Some researchers have even suggested differences in the mirror neuron system between men and women, backing up the oft-quoted (but little justified) assertion that females are blessed with greater empathetic skills than men.

Not only are our selves transient, mobile and destructible, but there seem to be more than one of them. The neuro-scientist Antonio Damasio says that our sentience can be divided into a 'core' self that reacts to stimuli, building a picture of the 'now' in the brain, and a more reflective 'extended' self, which relies on memories and building a picture of an anticipated future. Disturbances to the mechanisms that give rise to these selves can have profound consequences. Dementia can wreck the extended self, while brain injuries (such as that suffered by Phineas Gage) can disrupt the primary self, leaving memories intact but, in the words of Paul Broks, 'recalibrating the machineries of emotion and temperament'.

Insights into the nature of the conscious mind come from illness. People suffering from transient epileptic amnesia may lose their extended self and become a floating mass of awareness with no identity. People whose brains have been damaged by stroke may lose all sense of personal identity and yet be fully conscious and in some senses functioning.

We are not the same as we were a minute ago, in this sense science does have an answer to this question. We are not even the same from one bit of our brain to the other. Our selves are hugely defined by our memories and yet these can be as false as a childhood storybook.

But this does not mean we are really anywhere near solving the fundamental problem of explaining self-awareness. The more science probes the brain, the more it becomes clear that our 'folk science' ideas about self and identity will have to go out of the window. So might the whole concept of 'spooky' consciousness, which may have to go the same way as phlogiston and the philosopher's stone. Once it is understood how physical neurological processes generate a feeling of self-awareness, then the need to invoke this mysterious entity will disappear. The most obvious practical implication is that we may one day have to think rather differently about how we treat our criminals. The legal concept of lifetime responsibility may have to change.

Losing the soul sounds like a depressing thing, another victory for cold reductionism. But in another way it is liberating. We really can live for the moment, because we have no choice.

8

why are we all so fat... and does it really matter?

The current epidemic of obesity is one of the most extraordinary phenomena in the history of human health and wellbeing. If there is one reason why I am not the same person I was a month ago, it is because of the inevitable addition of a few unnecessary grams to my person. The reasons for it appear obvious and the solution equally so.

But in subtle ways, the plague of fatness defies logic and divides nutritionists, doctors and the overweight themselves. Like animal consciousness, this has become far more than a purely scientific question. It touches on morality and politics as well. The obesity crisis has spawned dozens of crackpot theories and fads, and made as many fortunes for those willing to exploit the misery of the overwhelmed hordes. Meanwhile our technology, globalized culture and media generate ideals of human perfection and beauty that would have made Helen of Troy blush and check for cellulite. We are fat people living in a world made for the thin.

The statistics (if of course you believe them) are simply mind-boggling. There are now more people alive eating themselves into an early grave (a billion overweight, one estimate said in 2006) than who do not have enough food to live (800 million suffering from malnutrition). In some parts of the world the seriously overweight comfortably outnumber the svelte and even the mildly chubby. America, with 30% of its population obese, is often cited (discounting island oddities like Samoa and Tonga) as being the global capital of fat. And the rest of the world is catching up.

In the UK, two thirds of all adults are now classed as overweight and a full quarter obese. In Scotland, a greater percentage of children now are obese even than in the US.

Most startlingly, child and adult obesity rates across a number of countries, including the US, Canada and much of Western Europe, have tripled or quadrupled in the past 30 years or so. One 2006 Department of Health survey predicted that by 2010 there will be 12 million obese adults and a million obese children in the UK. By the middle of the century, if current trends continue, nearly *everyone* in the US will be overweight and a majority of the population clinically obese. Obesity rates are soaring across Asia, South America and even Africa (although malnutrition, not overfeeding, remains that continent's biggest problem).

And with obesity comes a whole host of known problems and side-effects; some entirely predictable, some less so. Being very overweight has an obvious negative effect on quality of life. Serious conditions such as heart disease and diabetes are more common in the obese. Being hugely overweight is bad for you.

This tidal wave of fatness is affecting the very shape of our world. Everything has to be slightly bigger than it was even a quarter of a century ago. Bus, train and aircraft seats are having to accommodate our ample behinds. In areas where we can specify our own space we have supersized up. The cars of today are bloated whales compared to the sleek minnows of yesteryear. The average automobile on sale in Europe today weighs around half a tonne more than it did in 1977. Part of this comprises safety features: airbags, crumple zones and luxury gadgets of course, but a lot is to do with the sheer increase in size needed to swallow up our enormous bottoms. The US car market, after decades of downsizing, has now reverted to type. More than half of all vehicles being sold today are trucks – or SUVs – partly because when you weigh a fifth of a tonne and your teenage daughter weighs more than Mike Tyson you really do need a truck to carry you around. Our girth is making us drive bigger cars and is hence contributing to global warming.

The story of obesity looks, at first glance, to be a simple one. We are fat because we eat too much and exercise too little. We have become lazy and indolent. Being so fat, we are putting our health at terrible risk. Most government-funded experts warn of an obesity-related health crisis in years to come. And the solutions are equally simple. Eat less, eat better food and exercise more.

But there are some big fat problems with this simple thesis. For a start it is not clear that being fat is necessarily making us as unhealthy as we have been led to believe. The biology of obesity is turning out to be more complicated than was once thought; it may be far more complicated than simply calories in, calories out. And the solution, if some maverick scientists are right, may involve rather more than going on a diet.

The obesity epidemic certainly doesn't – so far – seem to be doing us as much harm as feared. In August 2006, a study conducted by the Mayo Clinic in Minnesota was published in *The Lancet*. The report was a *metastudy* – a study of published studies – and the findings took in results from a total of more than a quarter of a million patients looking at the link between weight and health. The statistics were simple and the findings pretty conclusive, so, unlike many counterintuitive 'studies' that make headlines, this is one that can probably be taken seriously.

The Mayo team found that people who were overweight and who had already been diagnosed with heart disease had significantly better survival rates than those classified as 'normal'. Even mildly obese people did better than those with the 'healthiest' body mass indices (BMIs), the standard catch-all assessment of weight and health. You actually had to be severely obese to do less well than someone in the normal weight band.

While the scientists took some pains to avoid being quoted as saying 'being a bit fat is actually pretty healthy', this was the unavoidable conclusion. In fact, what the findings highlighted the most was the probable unreliability of the BMI as a way of gauging anything useful about one's health. Dividing your weight in kilos by your height in metres squared yields a number which may or may not be as useful as your foot size when it comes to determining the likelihood of one living to a ripe and healthy old age.

Many elite athletes fall into the 'obese' range. The New Zealand rugby union player Jonah Lomu is probably one of the fittest men on the planet, yet he is officially obese, with a BMI of 32. Hollywood stars Brad Pitt and Russell Crowe are overweight and borderline obese, respectively, despite there being no evidence that either man is suffering from any obesity-related illnesses.

Weight *per se* may be a poor indicator of health, simply because heavily muscled individuals are relatively dense (producing high BMIs), whereas a bulk of fat is of a much lower density. Many doctors now prefer to use the simple waist measurement, which apparently correlates far better with predicted health outcomes. Even in the US, few people are extremely obese (BMIs of more than 40) so it is perhaps not surprising that the predicted health and life expectancy crisis has so far failed to materialize.

Morgan Spurlock, an American journalist, made a hugely successful movie in 2004 called *Supersize Me*, in which he lived on nothing but the products of one fast food firm, McDonald's, for several weeks. We were told that as a result this previously lean, healthy young man put on a lot of weight, became stressed, lost his sex drive and, most terrifyingly, started to suffer from incipient liver failure. This film is now cited as strong evidence that you get fat by eating fast food, and as a result you will also start to die.

In fact the media have been rather uncritical of *Supersize Me*. McDonald's is of course a global hate figure for many, and it makes an easy, if not lazy, target. When I spoke to one extremely eminent liver expert after the film was released, he voiced extreme surprise concerning the effects that Spurlock's diet had apparently had on his liver in such a short time.

In 2006, Spurlock's experiment was repeated, this time under controlled laboratory conditions, in Sweden. Fredrik Nyström from Linköping put 18 volunteers on a 'supersize' diet – not just McDonald's, but consisting of an instruction to double their daily calorific intake using junk food, and to avoid physical activity as much as possible (another aspect of Spurlock's regime).

The volunteers were poked and prodded throughout, and a new X-ray technique called DEXA (dual energy X-ray absorptiometry) was used to measure muscle, fat and bone density accurately. A barrage of liver tests and blood cholesterol levels were taken. The works, in short.

The results were extremely interesting. The volunteers, unsurprisingly, put on weight, but in wildly differing amounts. One, Adde Karimi, a nursing student, put on just 4.6 kilograms, and half of that was muscle. And this after a month on 6,600 kilocalories a day and virtually no exercise.

Additionally, his cholesterol levels actually dropped. Another volunteer put on 15% extra body weight in just two weeks. 'Some people are just more prone to obesity than others', Nyström told *New Scientist*, which reported the experiment in January 2007. Interestingly, none of the first batch of volunteers in the Swedish study suffered the elevated liver enzymes that caused Spurlock's doctor to make him quit his experiment, although some later volunteers did suffer this sign of liver damage.

The message from this experiment? It really is more complex than calories in, calories out. And the message that a high-calorie, fast food diet will *inevitably* make you ill is probably incorrect. On identical regimes, some people put on weight far more easily than others.

It has been clear for some time that the link between diet and cholesterol levels is tenuous, to say the least. But this experiment showed that we may have to throw away our preconceptions about this chemical; some of the Swedish volunteers had reduced cholesterol levels and even increased levels of 'good' high-density lipoprotein. And this is not the only *Spurlockesque* experiment performed recently. A batch of other experiments, some amateur stunts, others more scientific, appear to show that consuming fast food is not enough on its own to ruin your health and become fat.

The old fatties' excuse, 'It's my metabolism', really might have some credibility after all. The extra calories have to go somewhere, of course. Nyström suspects that 'naturally thin' people, if there are such things, simply burn off the excess energy as heat. He noticed that his initially thinnest volunteers complained a lot about being hot and sweaty when on the diet.

Interesting though these results are, it is undoubtedly the case that obesity must be linked to how much we eat. Fatness is hugely related to class and income, particularly in the West. All food has got hugely cheaper since the 1950s, and cheap food has become, relatively, cheapest of all. So while the price of high-quality fresh fruit and vegetables, well-reared meat and fish has fallen, the prices of cheap breads and cakes, candies and low-quality fast food that requires little or no preparation have fallen fastest of all. In the 1950s, when the hamburger chains began their relentless expansion across the US, a meal in a fast food restaurant cost around a working man's hourly wage. There are hamburgers now on sale in the US for as little as 39 cents, which would take about three minutes to earn on the minimum wage. Calories have never been so cheap at any place or time since we gave up hunter-gathering.

How we get our calories has changed too. If you walk around an American supermarket you will see nothing but 'lite' and 'lo-fat' products on the shelves. What you won't see is an admission, on much of the packaging, that a lot of this fat has been replaced (because it needs to be if this stuff is to remain edible) with something called high-fructose corn syrup (HFCS), which has replaced good old sucrose as a ubiquitous ingredient in American processed foods. Corn syrup, cheap and easy to produce in vast quantities, has been the subject of much media interest and has been dubbed 'the devil's candy'. It may play a role in the obesity crisis.

Its use in the US dates from the same time – the early 1980s – that the fat epidemic began to take off. But there is little evidence that this stuff is to blame. It is hard to explain, for instance, why eating HFCS should make you fatter than sucrose. Maybe it is a gloopy red herring.

Another simple factor must be the rise of the automobile. Gasoline is now cheaper, in real terms, than it has ever been, certainly in relation to wages. So are cars. It is now quite affordable for even quite poor people in the richest countries to drive everywhere, and most do.

Obesity rates in the US are among the highest in the sprawling cities of the south and southwest, where distances between homes and basic amenities are such (coupled with a simple lack of sidewalks and an often harsh climate) that getting everywhere by car is a necessity. In many of these places you simply cannot walk to the grocery store even if you want to. People are thinner in New York City than they are in urban Texas at least partly because it is impossible to park and mostly because everyone can and does walk everywhere.

Plus, of course, fast and (in particular) unhealthy-and-fast food is now the subject of multi-billion dollar marketing campaigns, often aimed specifically at hooking children into lifelong habits while they are at their most impressionable.

All in all, we are generally living a more sedentary lifestyle than ever before. British civilians were probably at their healthiest during and after the Second World War, when a combination of austere food rationing and petrol shortages led to the UK having probably the healthiest diet in its history, coupled with a mass, involuntary exercise programme involving long commutes by cycle and on foot.

The generation of Americans raised in the 1940s was in even finer fettle. Beef- and corn-fed, the strapping youths – farmboys and factory workers – sent over to Europe to fight the Germans seemed to the locals like supermen.

A hundred years ago the majority of workers worked with their hands. Now, muscle power is no longer in demand. It is hard to believe just how strenuous life was even four or five generations ago. As well as walking everywhere, keeping house involved hours of backbreaking drudgery; farm work was even harder, as the Industrial Revolution spawned countless millions of jobs in which brawn was far more important than brain. Nowadays, the growth of the service sector in all developed countries means that for most people exercise, if taken at all, has to be deliberately taken.

Millions of people in the West probably take no effective exercise at all, save the walk from their car to the office and vice versa. In the richest (and some of the fattest) countries parental paranoia has seen a sad generation of children hidden indoors in front of their computer screens when they should be outside playing, and driven everywhere, when they should be walking or cycling to meet their friends or to go to school.

But explaining the obesity epidemic cannot be this simple. We do indeed eat a little more and exercise a little less than, say, the generation born in the 1930s, but poor diets and sloth have not increased markedly since 1980 – when the obesity epidemic really started to take off. And the fat plague is really a staggeringly recent phenomenon. It is not even a whole generation old. According to the US government's Centers for Disease Control and Prevention:

... in 1995, obesity prevalence in each of the 50 US states was less than 20 percent. In 2000, 28 states had obesity prevalence rates less than 20 percent. In 2005, only 4 states had obesity prevalence rates less than 20 percent, while 17 states had prevalence rates equal to or greater than 25 percent, with 3 of those having prevalences equal to or greater than 30 percent (Louisiana, Mississippi, and West Virginia).

Car ownership and use in 1970s America was certainly a little less than today, but not hugely so. Diets may have been slightly different, but, surprisingly, Americans now consume less fat and much less meat than they did a generation ago (although they do consume rather more sugar, especially that fructose syrup).

While it is mildly challenging to scientific orthodoxy to state that some people may be able to eat far too much (and even be very fat) and yet remain healthy, it is utter heresy to suggest that some people may become obese because of an infectious agent. And yet the recent idea that you can catch fatness is proving surprisingly hard to dismiss.

The idea that you can catch obesity seems counterintuitive and absurd. So did the idea that you could catch stomach ulcers until Australian pathologist Barry Marshall drank a cocktail of *Helicobacter pylori* and won a Nobel Prize for proving that you could. Infectious agents are now suspected to the cause of a host of conditions, from schizophrenia to heart disease, that were once assumed to be the result of environment factors or genes. In 2001, a team from Johns Hopkins University found that people with schizophrenia were more likely than the general population to carry an activated version of a retrovirus called HERV-W in their DNA. Again, this is not to say that schizophrenia is an infection, but it may suggest that blaming the disease purely on genes or upbringing may be too simple.

And obesity? Some doctors, led by a charismatic Bombayan called Dr Nikhil Dhurandhar who now works at the Pennington Biomedical Research Center in Baton Rouge, Louisiana, think that infection may be a factor in the fat plague.

Dhurandhar had suspicions about a type of pathogen called an adenovirus; various strains are responsible for diseases like colds, diarrhoea and conjunctivitis. Animal experiments carried out in the early late 1990s and early 2000s showed that one strain, AD-36, could cause spectacular weight gain in infected marmoset monkeys. More evidence that viruses may play at least a part in unexplained obesity came with the finding that, in blood taken from 313 obese people and 92 lean people from Wisconsin, Florida and New York State, antibodies were present in just four of the lean subjects and 32% of the overweight ones.

To check that it is the virus making people fat, not that it is being fat that increases one's susceptibility to infection, he looked at the prevalence of three related adenoviruses – AD-2, 31 and 37 – and found no difference between the obese and non-obese populations.

The history of science is littered with spectacular claims that X causes mysterious Y, usually made by charismatic and highly-qualified people, that fade into nothing. Often what sinks such claims is the lack of a plausible mechanism, but Dr Dhurandhar has an answer. He has discovered that the virus appears to target the immature precursors to fat cells, altering their DNA and speeding up these cells' maturation.

This is not a theory shipwrecked on the wilder shores of implausibility. Dr Dhurandhar's work has been published in numerous peer-reviewed journals, including *the International Journal of Obesity and Obesity Research*. 'When I started out I guess my credibility rating was zero per cent', he says. 'Now it is maybe 60 or 70'. Most mainstream nutritionists are still pretty dubious, but some, such as Dr Iain Bloom, a metabolic medicine specialist at Aberdeen University in the UK, are cautiously supportive.

So this is still far from the mainstream, but a long way from being bonkers. Interestingly, another of Dr Dhurandhar's findings suggests that adenovirus-triggered obesity may be associated with increased sensitivity to insulin, in turn suggesting that people who become overweight may be actually less susceptible to diabetes than the general population. Depressingly, if adenoviruses do indeed turn out to be a co-factor (probably acting in concert with some sort of genetic susceptibility) to one 'strain' of obesity, the chances are that, like any viral infection, it will turn out to be essentially incurable.

The adenovirus may not be the only micro-organism making us fat. In December 2006, *Nature*[1] published a report by Jeffrey Gordon and colleagues from the Washington University School of Medicine in St Louis, Missouri, showing that the gut flora of a subset of obese people (and obese mice) differ subtly from the microbes found in lean individuals. Specifically, the ratios of the bacterial groups known as *Firmicutes* and *Bacteriodetes* were different. Gordon's team surmises that the bacteria in the 'obese' guts enable their hosts to metabolize calories more efficiently, leaving fewer nutrients to go to waste in faeces.

It is thus plausible that some sort of symbiotic mechanism has evolved where 'friendly' gut bacteria enable us to get the most out of every mouthful (a good thing in normal circumstances, but a bad one where food supplies are unlimited).

It is very early days. It may well turn out that the 'microbe that causes obesity' will disappear along with the Snark. But there is now quite a lot of evidence that the story of our expanding waistlines may be a little more complicated than a simple tale of sloth, greed and increasing prosperity. It is certainly the case that when everyone starves, none are fat. But it is also certainly the case that when food is plentiful some people remain lean, even when they appear to consume a great deal. Most challenges to scientific orthodoxy turn out to be diversions.

A few – a very few – do have something to them; maybe the infection–obesity link is one such. There is certainly a lot that needs to be explained about why we are all becoming so fat – the mysterious maintaining of general health levels and life expectancy, the speed at which the phenomenon has grown and in particular the interesting microbiological findings which suggest that as well as diet, exercise and genes, sheer bad luck may play a role in expanding your waistline. Fat is a feminist issue. It is also a scientific one, and the science is turning out to be not as straightforward as we once thought.

Reference

1 Ley, R. E., Turnbaugh, P. J., Klein, S. and Gordon, J. I. (2006) Microbial ecology: human gut microbes associated with obesity. *Nature*, **444**, 1022–3.

9

can we really be sure the paranormal is bunkum?

Like most people who consider themselves to be rational beings, I have a long hate list of tiresome beliefs, notions and lifestyles which I reckon consign a person to the dark side. This list includes the following:

- all religions, whether organized or utterly chaotic
- astrology, spoon-bending and other alleged demonstrations of psychic powers such as telepathy
- any manifestation of the New Age, including talk of chakras, channelling, crystals and chanting
- the 'wisdom of the East' and the 'wisdom of the Ancients'
- the healing power of whale music
- aromatherapy, rebirthing and reincarnation
- alternative medicine, especially homeopathy and really especially *homoeopathy*
- essential oils
- anything Ayurvedic or involving gurus
- ghosts, fairies and, far worse, faeries

Collectively all this stuff brings me out in a rash. Indeed, I suspect that the barstool question 'What star sign are you?' is an excellent Darwinian adaptation hardwired into the mating strategies of the deluded to provide a signal to sensible folk to keep clear and protect the gene pool.

In my world view (and in the world view of most people I know and love who are all, of course, sensible right-thinking people like me) this is all gibberish, wishful thinking by people who are not interested in finding out how wonderful the world really is and who wish instead to replace it with a garish, Disneyfied version where there are fairies (or faeries) at the bottom of every garden and a big, kind man in the sky to watch out for our every move.

I was especially pleased to learn from a psychology professor once that a belief in things like astrology and mysticism is hugely correlated with a highly conservative right-wing outlook on life generally. Great, I thought. These kooks are not only talking gibberish, they are a bunch of Nazis as well. It's therefore more than all right to hate them, it is almost a duty.

Here is another list, this time of things I do believe in:

- huge stars which collapse in on their own weight and at whose centres there are maybe portals to another universe
- the possibility that an infinite number of parallel universes exist, each containing every possible permutation in the history of time. I believe it is just about possible that out there, there is not only a universe where Al Gore became President of the US in 2000, but an unhappier place where Hitler won the Second World War.
- objects like electrons and maybe even whole atoms can be in two places at once
- when you put a stopwatch on an airliner and fly it over the Atlantic the act of accelerating this object to a few hundred miles per hour will make it run very slightly slow
- the Universe began in a stupendous explosion of space, matter and perhaps time as well, and we have managed to find a date, around 13.7 billion years ago, when this event occurred
- the Universe is full of a strange invisible substance which completely fails to make its presence felt save through its gravitational attraction to ordinary matter. I am quite prepared also to believe in an even more mysterious, monstrous dark force that looks like it might one day rip everything apart

So what is it that differentiates the first set of beliefs from the second? What makes the second lot 'scientific' and the first lot 'mumbo jumbo'? Why is belief in homeopathy silly and yet belief in string theory completely sensible and mainstream? Why do you get Nobel Prizes for working in one set of these fields and derision for working in the other?

The answer gets to the heart of what science *is*. The rejection of 'flaky' beliefs like chakras and homeopathy is not because these things are intrinsically strange, or even spooky. After all, quantum physicists have performed experiments which have shown that two electrons can send 'messages' to each other thousands of times faster than light. Any explanation for this 'entanglement' involves hypotheses like sending messages back in time, which is far, far spookier than homeopathy.

No, science does not reject certain beliefs because they sound flaky, but because they have been examined by experiment and found wanting. The scientific method says that you have an idea, and test whether it is true. *Belief* – except in the veracity of this method – does not, or should not, come into it. And time and time again, when science has tried to verify things like homeopathy or the existence of telepathy, it has failed. The 'proof', if there is any, for the phenomena on the first list depends hugely upon anecdote. And anecdotal evidence, while not always entirely worthless, is generally the enemy of reason.

But we must be careful here. It is tempting to dismiss a whole set of beliefs – indeed a whole belief system – purely on the basis of prejudice rather than on evidence. And it is far too easy to link one set of beliefs (for which there is no evidence) with another (for which there might be) simply because they sound a bit similar and tend to have the same devotees. People who 'believe' in crystal therapies and chakras often also believe in acupuncture and telepathy. And while there is zero evidence that the first two are real, there is quite a lot of evidence that acupuncture 'works', and some evidence for telepathy.

Believers in the paranormal often point out that science is quite prepared to believe in stuff that is not only spooky (like entanglement) but also stuff for which there is very little experimental evidence.

They have a point. Take string theory, the idea that at its heart the Universe is composed of countless tiny vibrating threads made of, perhaps, space–time. Each is tuned to a different harmony, creating the electrons, quarks, neutrinos and so on of which we are made. It is a beautiful theory, and, at least in its very broadest concepts, quite a simple one, and if there is one thing that science has taught us it is that the simplest answer is very often the right one. But is it right?

Mathematically, string theory (I am told) pretty well has all its ducks in a row. When I met Lisa Randall, the Harvard University physicist and evangelist for string theory as the best possible candidate we have for a prototype theory of everything, she certainly impressed me with her passion. People like Randall live in a mental world upon which we can only gaze in awe. Who on earth are we to doubt them?

But some people with far more mental equipment at their disposal than I do doubt all this. String theory has, so far, absolutely no observational or experimental evidence to back it up. Already there is something of an anti-string theory backlash. The problem with string theory, say the sceptics, is that it is fundamentally *untestable*, and that makes it fundamentally Not Science. Because to expose particles on this scale and study them we will need to build atom-smashing machines two or three orders of magnitude more powerful than anything we have mooted so far.

And the scepticism about the New Weird Science doesn't stop there. Parallel universes are an elegant solution to two large scientific dilemmas: the resolution of quantum events and the problem of explaining why the Universe seems to be so finely tuned for life. And yet, like those strings, we have absolutely no empirical evidence that there are universes out there where Hitler won the war or where another you is reading *Astrology for Cats* right now.

Part of the reason that string theory and, say, homeopathy have been put in separate boxes is of course the personalities involved. The people working on string theory, like the people working on dark matter and the people trying to fathom the nature and possible cause of the Big Bang are proper scientists, proper people in fact, clearly highly intelligent and with years of training behind them.

They submit their findings to respected journals where their peers mercilessly rip their work to shreds, trying to find fault and any evidence of error or fraud. Their hunches, hypotheses and theories are testable and fallible and their experiments repeatable, and that is what makes them proper scientists, not charlatans.

Some of these people become stars, write best-selling books and make a great deal of money, but most do not. Some are impossible egoists every bit as queeny as the worst showbiz diva. But again, most are not. The majority of top-end scientists I have met, even Nobel Prize winners, are surprisingly unassuming people, and perhaps a majority of them find the fame which may be unwittingly thrust upon them embarrassing and hard to deal with.

Compare these people with the other lot. They often wear silly clothes and spout unfathomable gibberish, and their most successful proponents seem to worship both fame and money a great deal. Little training is required to set oneself up as an astrologer, faith healer or spoon-bender, just a 'gift', some charm, a thick skin and a snappy personality. These people often react very badly when asked to put their findings or qualifications up for serious scrutiny, and they often reach for their lawyers at the suggestion that they may be mistaken in their beliefs.

The work of these people is helped greatly by the strange, modern climate of scepticism, indeed cynicism, of all things scientific, a rejection of 'modern thinking' and indeed, implicitly (although it is never quite put like this) of the whole Enlightenment project.

Finally, there is the unalterable fact that, historically, much of even the edgiest science turns out to be right. It sounds as silly to believe that a pocket watch on an airliner runs slow as it does to believe that the position of the planet Neptune when you were born may influence your career and choice of partner, but the fact is that we can measure the former very accurately with atomic timepieces, and it does indeed turn out to be the case that fast-moving clocks runs slow. And it also turns out that the position of Neptune has, statistically, no influence at all on the course of your life.

Similarly, the notion of dark matter sounds absurd until you realize that with some very elaborate and expensive telescopery and computery, you can actually see the gravitational shadow of this stuff writ large in the heavens. No one has ever got a paper in *Nature* based on their spoon bending powers or their ability to predict the future.

And yet. There is a danger here that we are creating a false dichotomy, an unnecessary barrier between the logical and the absurd that is really a barrier between two different mindsets than between the real and the unreal. Strange, impossible, weird and even spooky things are OK as long as they are 'scientific', but not if they are just spooky. Quantum action at a distance is all right; ghosts are not. Quantum consciousness is worthy of debate, telepathy is beyond the pale. NASA employs scientists to ponder the existence of microbes on Mars, yet to believe seriously in UFOs is to stray well into nutterdom and certainly will count against you if you are going for a job with the Agency.

While we can all agree that crystal therapy and channelling are almost certainly nonsense of the first order, what about hypnotism and acupuncture? Can we really be so sure that all this stuff, the stuff of the loonies, the credulous and the fraudulent, should be thrown away with the mystic and astrological bathwater?

Do we need to think a bit more carefully about exactly where we place our Great Wall between the rational and the absurd and even allow for the fact that sometimes that wall may need to be given a few gaps?

Fundamentalists would say that giving airtime to any of this stuff is an abomination. I remember hearing a very distinguished British thinker saying on the radio a few years ago that even if it could be shown, beyond all reasonable doubt, that telepathy was real, he would still want nothing whatsoever to do with it. Such a discovery, he said, would be trivial and unimportant. It would tell us little that we didn't already know about the brain and the mind, how the world works and how we fit into it.

Well, I am sorry, but this will not do. As the Nobel Prize-winning physicist and paranormal sympathist Brian Josephson says, this is a 'pathological disbelief... a statement which says "even if it were true I wouldn't believe it"'.

It is certainly the case that if it could be shown, for example, that telepathy works it really would change an awful lot. If we discovered that brains are able to communicate, through empty space, directly and without the intermediary of spoken language, then this alone would tell us a great deal about human consciousness, the mind, and the transmission of information. Obviously I have no idea how telepathy works, if it does (which I doubt), but that is not the point. Maybe it would involve some sort of quantum spookery, maybe some sort of electrical field.

According to Richard Wiseman, a British psychologist who has spent many years studying and commenting upon parapsychology, the discovery that any of this stuff is real would be hugely important:

It would not be a small change to our scientific model of the world if astrology, ESP or ghosts were genuine. It would be a radical shift. That is why these topics make many scientists instantly say that these things cannot be true. We must remember that about half the public believe in these things and so they are deserving of investigation from that perspective alone.

Science is in the business of testing ideas, torturing them, wringing every last possible anomaly out and hanging them to dry. If you prove that something is right, that isn't good enough. You have to show everyone else what you have done and they must repeat your experiments and get the same results. Only then has knowledge advanced.

Can any of this rigour be applied to the paranormal? Well yes, actually. Parapsychology is the name given to a group of (so far) hypothetical phenomena which include extrasensory perception (ESP), telepathy, clairvoyance, precognition, remote viewing, telekinesis, psychic healing and morphic fields.

What these phenomena have in common is that they are, at least in principle, testable. And since the 1890s there have indeed been concerted efforts made to find out whether these are real, in controlled conditions, in the lab. Some of these experiments, where people sit in sealed rooms and try to transmit images on cards – circles, squares, wavy lines and so on – to another volunteer in another sealed room, have become quite famous.

So far the results have been rather muddled. Some individual studies have shown some sort of statistically significant effect (i.e. the 'receivers' were getting the 'right' answer far more often than would be expected simply by chance), but the sceptics point out that if you do a 'study of studies', if you look at dozens or hundreds of individual attempts to 'find psi' the interesting results disappear in a puff.

It is fair to say that no investigation into telepathy has yet produced results which have convinced the scientific mainstream that there is any kind of interesting effect going on here. The parapsychologists meanwhile maintain that such metastudies instead reinforce the certainty that something interesting is going on.

Perhaps the most 'successful' of the ESP tests are the 'Ganzfeld' experiments, which were first conducted in the 1970s. Volunteer 'receivers' are put into a state of near sensory deprivation, in a soundproof room with translucent spheres placed over their eyes, bathed in red light and with white noise played through headphones. The idea is to create a 'changeless sensory experience', fully open to any telepathic signals, should they be there.

There are several variations in the methodology, but basically the receiver is asked to rank a series of images in terms of how well they correspond to 'signals' sent by a 'transmitter' in a sealed room. According to experimenters, overall Ganzfeld trials have shown the existence of psi effects beyond all reasonable doubt – one figure quoted is that you would expect results such as have been obtained to have occurred by chance only one in 29 quintillion times.

Since then, however, meta-analyses of Ganzfeld experiments, by Richard Wiseman and others, have apparently shown no such effect. There have been allegations of experimental error and irregularities. Perhaps most importantly, the assumption that any statistical anomaly must be due to a mysterious psi effect has been challenged; maybe telepathy is at work, but maybe there is a hitherto unsuspected error in the experimental protocol.

Experiments are done, and results argued over, and parapsychology as a whole goes in and out of fashion. A hundred years ago it was reasonably respectable, with even Darwin's great protégé Alfred Russel Wallace having a thorough dabble. Then, as the 20th century dawned, an extreme, ratio-nalist mindset took hold and any sort of psi research became seriously beyond the pale.

And after that, came the ESP experiments and now, finally, the new scepticism. It is true that several respectable universi-ties now have departments and research bodies devoted to parapsychology, but despite this the whole field is still tainted by its associations with the charlatans and frauds who perpe-trated the spiritualist movements of Victorian times. To some, this has led to irrational prejudice. Brian Josephson has accused mainstream journals such as *Nature* and *Science* of effectively censoring any papers on telepathy, telekinesis and so on. Put it this way: expressing an interest in psi effects won't see the research grants flooding in.

Scientists have a natural inclination to distrust ideas that are outside their canon of knowledge. There is a sense that para-psychology is partly 'owned' by people outside the traditional field of psychology, for instance. This may explain a remark-able and enlightening finding that came in 1979, when a survey of more than a thousand American college professors found that a majority (55%) of natural scientists, a large majority (66% of social scientists) and a huge majority (77%) of arts professors were prepared to accept that ESP was at least a possibility worth studying. The only group which expressed extreme scepticism were the psychologists (only 34%), and a similar number said that ESP was an *impossibility*, a view taken by only one in 50 scientists generally.

In a paper published in 1994 in the journal *Psychological Bul-letin*, entitled 'Does psi exist? Replicable evidence for an an-omalous process of information transfer', Daryl Bem and Charles Honorton had this to say about these figures:

> We psychologists are probably more sceptical about psi
> for several reasons. First, we believe that extraordinary

claims require extraordinary proof. And although our colleagues from other disciplines would probably agree with this dictum, we are more likely to be familiar with the methodological and statistical requirements for sustaining such claims, as well as with previous claims that failed either to meet those requirements or to survive the test of successful replication. Even for ordinary claims, our conventional statistical criteria are conservative. The sacred p = .05 threshold is a constant reminder that it is far more sinful to assert that an effect exists when it does not (the Type I error) than to assert that an effect does not exist when it does (the Type II error).

This amounts to a defence of psychological scepticism. Richard Wiseman adds:

Psychologists have carried out lots of work showing that people are often driven by their beliefs when they evaluate evidence, rather than being more rational. Also, they obviously carry out work with people, rather than with chemicals, and so are used to people cheating, not telling the whole truth and so on. As such, I think they are more aware than most of how evidence for an effect may be due to human deception and self-deception.

In other words, psychologists work in a world where people lie a lot. Physicists do not. This makes physicists a bit more gullible.

In their paper 'Biological utilization of quantum nonlocality', published in *Foundations of Physics* in 1991, Brian Josephson and Fotini Pallikari-Viras cautiously floated the idea of the phenomenon of quantum entanglement, which Einstein famously dismissed as 'spooky action at a distance' as a possible mechanism for telepathy.

It is no surprise that the known strangenesses of the quantum world have often been cited as a possible (perhaps the *only* possible) explanation for various paranormal effects from ESP to the phenomenon of consciousness itself. The Oxford physicist Roger Penrose has suggested that microscopic structures inside the brain, microtubules (which are in fact found in all cells), may be able to make use of quantum effects to produce the non-deterministic effects of self-awareness and free will, a view dismissed by many of his peers as twaddle.

In brief, Josephson and Pallikari-Viras say that it is not impossible that the existence of 'remote influences' suggested by quantum theory (where, say, the quantum state of an object like an electron or photon, say its spin, or polarization, may correlate over arbitrarily large distances after they have been split apart) may indicate that the same effect could lie behind the direct connection of minds (telepathy) and between mind and matter (telekinesis).

As to an actual mechanism, the authors accept that invoking quantum effects in a macroscopic structure like the brain is stretching credulity, but that it is quite plausible that during the long evolution of life on Earth natural selection has, in effect' tamed the quantum world to use its properties for its own purpose.

A grand interconnectedness between *all* life-forms is invoked, a sort of super-Jungian megaconsciousness. Today, Brian Josephson says the idea that we will have to 'throw away science' if we accept the reality of some psychic phenomena is:

> ... a woolly argument, nonsense. Fundamental physics may have to change a bit to include the mind but it is not true to say it would all have to go. When new findings come along in science it is rarely the case that all previous beliefs have to be overturned.

But the question remains: why should we take this seriously? After all, what is the difference between telepathy and the tooth fairy? It may be hard or even impossible to prove that these phenomena are not real, but what is the point of wasting time, energy and money investigating things which are probably marginal at best and very probably no more than a figment of our collective imaginations?

Well, for a start, even if we never do uncover evidence of psychic ability, by carrying out experiments into 'ESP' we are quite likely to find out a lot of interesting things about the psychology of the self and of deception. That alone makes this work worthwhile.

More fundamentally, parapsychology is a 'real' phenomenon if only in the way that so many people perceive it as so. Some sort of voiceless communication has been reported by human societies across all cultures and apparently across all times in history. Most cultures report instances where individuals are able to make contact with other individuals instantaneously and across great distances.

I think the interesting thing about this is that although the telepathy experience is quite common, it is by no means universal and when it is reported it seems to be a pretty marginal effect. In a way, it is more easy to dismiss beliefs such as the afterlife and the various deities, simply because such beliefs are so universal (and hence accepted and unquestioned).

Simple psi effects seem to be rare, and have always been thought of as something rather special and probably quite dubious. They also seem to be free of political, religious or emotional overtones. For what it is worth, I do not believe that telepathy and other related psi effects are real, or at least I do not believe that I have been shown anything to convince me that they are real, but I have no reason to believe with any certainty that they are not.

What about the rest of the paranormal? Well, it is possible to construct a sort of 'league of looniness', with the most plausible bits at the top and the most foam-flecked reaches at the bottom. At the top I suppose would be what the high priest of rationalism Richard Dawkins has recently dubbed the 'perinormal'. Here we find hypnosis and maybe acupuncture, both now, following solid clinical trials, largely accepted to be real, although mysterious, phenomena. Would he include telepathy? The sneer could be heard down the telephone line. 'Almost certainly not'.

Then the aforementioned classic psi phenomena – telepathy, remote seeing, perhaps telekinesis. Evidence for these being real is debated and hugely controversial. But compared to the next lot this is practically Newtonian physics.

Enter stage left here a curious, mostly North American, phenomenon called Intercessory Prayer (IP). This is faith healing, an unholy fusion of parapsychology, mysticism and traditional religion. In IP studies the effect on sick people of volunteers praying to God for their recovery is measured. (It never seems to be the case, puzzlingly, that the volunteers are asked to pray for a worsening in the patients' condition, although in the interests of scientific correctness surely this should be so.)

Several IP studies have been published showing an effect. For example, in 2001 Leonard Leibovici of the Rabin Medical Center in Israel had a paper published in the *British Medical Journal* in which he claimed that a group of patients with blood infections did slightly (but statistically significantly slightly) better than those who were not prayed for. Studies in the US have 'shown' small but significant effects on patients recovering after heart attacks and surgery.

IP studies are naturally hugely controversial. Why, many scientists argue, should money – sometimes public money – be used to fund such a flaky area of research and one so culturally specific? The idea of IP raises shudders outside the US Protestant heartland. And a lot of religious people are unhappy also; the idea that their God would choose to intervene to help some people and not others simply on the basis of a medical trial seems to undermine any commonly held views of an all-loving and just deity.

UFOs come next. Almost, but not entirely, implausible, the idea that Earth is being visited by alien spacecraft probably belongs in the same groupthink as IP and the kookier end of ESP. The arguments for and against UFOs are well worn, and not worth repeating here, except the one which says that if it were definitely and provably the case that no alien had ever visited Earth in a flying saucer (which it never could be) it would also most definitely be the case that once humanity had come up with the notion of aliens then flying saucers would sooner or later be seen.

Homeopathy? Nah. You can do double-blind trials – they *have* done double blind trials – and there is no effect (save perhaps a rather interesting placebo effect).

Reincarnation? What is the point? We are now on the ever-steepening and slippery slope that leads down into the intellectual dark side. 'Proper' religion probably belongs in its own category, perhaps a rival league, a bit like the two rival codes in rugby football.

Finally, it seems that a line must be drawn. Not a solid one, but a broken and permeable barrier between the acceptable and unacceptable. Richard Dawkins is probably being too fierce here, but his idea is a good one. Accept, grudgingly and with a certain amount of kicking and screaming, paranormal (all right: 'perinormal') phenomena into the scientific fold if and when there is overwhelming evidence that there is something here worth studying.

The paranormal is, probably, bunkum. Most of it, most of the time. But around the fringes it is just possible that science is starting to investigate something that is as hugely and deeply interesting as the wildest phenomena in the new physics and the new cosmology. If we are prepared to believe in dark matter, multidimensional hyperspace, dark energy and naked singularities before breakfast, I don't think a little telepathy should be too hard to swallow.

10

what is reality, really?

This is not a question about the meaning of life. That is whatever you decide to make it, and is a topic for discussion in the bar, not a book about science. Nor is this a purely metaphysical question, although it covers areas that have traditionally been the domain of philosophers. Instead, this is a question about the true nature of the Universe. At its heart is the ultimate, and for now totally unanswerable, question, why should there be anything here at all? As the physicist Stephen Hawking wrote, 'What is it that breathes fire into the equations? Why does the Universe go to all the bother of existing?'.

When scientists talk about reality they talk about tangible things – atoms and molecules, particles and radiation. But of course this is only the reality. Whether directly, through our senses, or indirectly, through our machines, we construct a picture of reality that resides not out in the stars and galaxies but within our heads.

The old solipsistic chestnut about the world possibly being a figment of our imagination cannot ever be dismissed out of hand. Neither, as we shall see, can the idea that the world, including ourselves, is a figment of someone *else's* imagination. That said, the fact that we have managed to formulate physical laws which correspond so exactly with what we observe suggests that while 'reality' may be what we perceive, we are perceiving something that is very concrete indeed.

But there is a lot we don't know about the ultimate nature of the Universe. For a start, what is its ultimate cause? Twenty years ago cosmologists stated flatly that the answer was simply 'the Big Bang' and left it at that, but now scientists are starting to realize that this is not good enough. What was the bang exactly? Why did it bang and what happened before?

We do not know if the laws that govern our Universe are arbitrary, or whether they are they are the only laws that there could be. Could, for example, the size of the gravitational constant just as easily have been double or half the value that we see? Or is it the case that there is a deep, underlying logic underpinning the laws of physics, like the foundations of a house, that dictates that there is only one possible way that a universe, if you are going to have a universe, can organize itself. And if so, where did these laws come from?

Perhaps the hardest question in physics, and one to which science has absolutely no answer, is that posed by Hawking; my understanding is that it can be summarised as:

> Is it the case that if there is a set of ultimate laws/logical propositions that underpin everything, do these laws in their nature demand not only the existence of the Universe but of themselves as well?

In other words, is it the case that it is impossible for there to be *nothing*? And if not, does the fact that there clearly is *not* nothing *mean* anything?

Finally, we can ask, is what we see really what we think we see? Humans have, ever since they started thinking about the world and its nature, come up with a number of outlandish folk cosmologies. The Earth sitting on the back of a turtle. The Earth as a disc floating in an infinite sea. The sky as a dome through which the lights of heaven are visible as the pinpricks of brilliance we call the stars.

Now we think of the Universe as a vast, 92 billion light year diameter sphere of expanding space–time driven by a mysterious dark force field that we do not understand and populated mostly by a ghostly form of matter that we cannot see and cannot feel. Is this any less strange than those old folk cosmologies? And is this the whole picture? Or is 'our' Universe simply a tiny mote on the back of a far vaster, far grander, appendage?

We do have some answers to these questions. Or at least, some ideas. The concept of the *multiverse*, a vast assemblage of universes, has become very popular in physics. By assuming a huge or even infinite number of parallel realities we can explain some of the oddities of the world we see around us, most notably the strange way in which the Universe seems to be so finely tuned so as to allow us to live in it.

But it gets even weirder than that. There are some left-field but logically quite respectable theories that state that just about nothing that we believe about reality to be true is in fact the case. The Universe, in these cosmologies, may be a setup, a creation not of a god or gods but of machine intelligences living in a world we could never see or fathom.

A popular solution to the initial cause problem, and indeed all the questions we have about the nature of the Universe, is of course God. Across most of the world and certainly for a vast majority of people the existence of some sort of deity forms a perfectly acceptable bookend to all chains of inquiry about themselves and the world in which they live. It is certainly the case too that even in our so-called secular age many scientists continue to believe in God.

Most scientists do not, any more, consider God to be a rational solution to the question of 'How did the Universe come into existence?' Old-fashioned Creationism, and its bowdlerized cousin, Intelligent Design, is the preserve mostly of Christian fundamentalists in the US, although there are worrying signs that it is making something of a comeback in Europe as well. If the answer *is* God then we might as well all go home. So let's move swiftly on.

Modern cosmology is of course on the case when it comes to these big questions, but just how far have we come from the days of flat earths and backs-of-turtles? As to the origin of the Universe, we now have a rather impressive model, the Big Bang, which seems to explain a awful lot, for instance the existence and observed spectrum of the Cosmic Microwave Background radiation (a dull glow, three degrees above absolute zero, that permeates the cosmos and is thought to be the dying embers from the Big Bang itself), the observed expansion of the Universe and the relative amounts of hydrogen and helium present.

But the Big Bang model is incomplete and there are many gaps. One important gap, albeit not with the model itself, is the general conceptual misunderstanding of the Bang as a gigantic explosion which threw vast quantities of shrapnel blasting into space, which later became the stars and galaxies. It is not entirely clear what happened, to say the least, but it is clear that the gigantic expansion of the universe that took place in the first millisecond after the Big Bang was an expansion of space–time itself, carrying the matter and energy embedded within it. It is better perhaps to imagine the Big Bang not as an explosion but more as the blowing up of a balloon.

But there are other, more serious problems, as even the Bang's most enthusiastic defenders will concede. For example, as we peer further away from the Earth we see galaxies as they were long ago. The light from very furthest objects that we can see left on its journey to Earth very shortly after the Bang, which is thought to have taken place 13.7 billion years ago.

These very distant, very early galaxies are only a few hundred million years old, as we observe them, and should therefore be packed with very young, immature stars (our star, the Sun, is more than 4.6 billion years old). And yet many appear not to be: some of these very young, very distant galaxies look like mature galaxies full of 'old' stars. Then there is the fact that some of the stars we observe seem to be 'older' than the Universe itself.

The evidence for this is far from clear-cut and is highly controversial, but it is worth pointing out just to show that the Big Bang is not a completely accepted model in the way that evolution, say, is accepted by biology. Even the fact that what the Big Bang created is so mysterious is problematic. The fact that the Universe is totally dominated by dark matter and dark energy – both utterly mysterious – is, as Dr Bob Nichol, an observational cosmologist at the University of Portsmouth in the UK, puts it 'a mystery and an embarrassment'.

What happened 'before' the Big Bang used to be seen as a pointless question, as it was considered that both space and time were created during the Big Bang; to talk of a 'before', therefore, is meaningless.

But this view has been challenged, most notably by the Cambridge theoretician Neil Turok and colleagues, whose 'ekpyrotic' or cyclic universe describes a bang that is not a real bang at all but due to events happening in a higher-dimensional space. The ekpyrotic model does not contradict the Big Bang, but sets 'our' bang in the context of a much greater ensemble of events and, crucially, no longer states that it is meaningless to talk of a precursor to the Big Bang.

The theory, in essence, states that 'our' universe 'floats' on a three-dimensional 'brane' which moves through higher-dimensional space. The Big Bang, for which we have so much evidence today, was an event caused when 'our' brane, after a period of contraction, collided with another, generating a great deal of matter and radiation.

Learning about the beginning of our Universe is proving to be expensive. Ordinary telescopes are good at peering back maybe halfway to the start of time, but even to get this far you need huge machines and the sort of computerized optics technology that has only been possible in the past decade or so. But to go back to the beginning, to peer literally into the mists of time, requires telescopes that can see in high radio frequencies in the millimetre range.

High in the Chilean Andes, at a literally breathtaking altitude of 5400 metres above sea level, the world's costliest ground-based telescope is under construction: the Atacama Large Millimetre Array (ALMA). Work has begun on constructing around 60 large dishes which, when the array is complete, will allow astronomers to see right back to the very early days of star and galaxy formation in the Universe, right back to the first few hundred million years. The young Universe is still something of a mystery. We don't understand, for example, how dark matter and ordinary matter interacted to form the earliest galaxies, and how this interaction operated through time to create the Universe we see around us today.

Insights into the very early Universe will also come from particle accelerators. As well as searching for dark matter particles, the collisions that will take place in the Large Hadron Collider in CERN will generate energies on a level similar to those seen in the Big Bang. It is an extraordinary thought – in a tunnel underneath the Franco-Swiss border it is possible to recreate events which took place nearly 14 billion years ago.

One of the big problems for a non-cosmologist is to work out exactly what the professionals are describing when they are talking about the 'universe' (and should that word have a capital 'U' or not?). Firstly there is the observable universe, which is the sphere of space surrounding the observer (in our case Earth) containing all places close enough for us to observe them. This means that the observable universe must be small enough to allow a light beam emitted by any object to arrive at Earth in less time than the (absolutely finite) time allowed since the Big Bang. This means that the observable universe, or Universe, while large, is most definitely finite; it has been calculated to have a radius of about 46.5 billion light years (the 'edge' of the universe is thus about 444,400,000,000,000,000,000,000 kilometres away), giving a volume of 3.4×10^{71} cubic kilometres. (This figure is not the same as perhaps would be inferred from the 13.7 billion year age figure; the reason the observable universe is not 27.4 billion light years across is because the warping of space–time allows light to apparently break its speed limit.) There are perhaps 80–200 billion galaxies in this sphere, called the Hubble Volume, and maybe a hundred billion stars in each, so the total number of stars runs well into the quadrillions. The Hubble Volume is expanding, by definition, at a rate of one light year per annum. It is big, impressive and contains an awful lot of stuff, but compared to everything – the 'true' universe – it is probably a gnat on the back of an elephant.

For a start it is an observable universe centred upon the Earth. No one believes that the Earth is at the centre of the Universe any more than they believe that the Earth is at the centre of the Solar System or that our planet is flat. The term 'observable universe' is a useful construct as a way of describing the maximum space in which things and events may be causally connected to us. Of course, observers in other parts of the Universe will each have their own 'sphere of influence'. The total number of galaxies even in 'our' region of space–time may be hugely larger than the number we can theoretically 'see' (the world 'observable' is theoretical and does not assume anything about telescope technology now or in the future). But even that may not be enough.

Just a few years ago the idea that the universe might be composed of a vast ensemble of unknowable parallel realities was pure science fiction. Now, some quite sane physicists postulate that to explain away the awkward fact that our Universe seems to be finely tuned for not only life, but for our sort of life, and that this is stupendously improbable, it is probably best all round if we assume that the Universe we think we live in is merely a tiny facet on an infinitely grander, huger diamond of creation altogether. In this *multiverse*, or *megaverse*, anything is possible, everything equally likely, and everything equally unlikely. The multiverse deals nicely with the so-called anthropic problem.

There are several variants of the multiverse hypothesis. One, the many-worlds interpretation of quantum mechanics, supposes that to resolve the apparent paradoxes of quantum theory every single possible state following a quantum event does indeed occur in its own universe. The main rival theory, the Copenhagen interpretation, states, in essence, that only one outcome is possible, arising as the 'wave function' collapses into one particular state (for example an electron being over there, or right here). Both interpretations are to an extent spooky. The multiverse hypothesis invokes the existence of billions of parallel realities. The Copenhagen interpretation seems to give a special role to the observer. Einstein objected saying, 'do you really think the Moon isn't there if you aren't looking at it?'.

There are other explanations for the multiverse. The miniverses may exist in an infinite or near-infinite volume of space, expanded to a preposterous degree by the cosmic inflation that took place since the Big Bang. In this multiverse, the brute-force multiverse, all possible outcomes are covered by the infinite nature of objects, including an infinite number of duplications (in this multiverse there is an identical copy of you, sitting reading an identical copy of this book in an identical room/bus/plane etc. It lies, on average, $10^{10^{29}}$ metres away. Then there is the multiverse proposed by the Russian-American cosmologist Andrei Linde, who has proposed a gargantuan ensemble of universes, each formed by an expanding bubble of space–time budding off from the others. In Linde's cosmology, our 'big' bang was just one of an infinite number of little bangs. Finally, individual elements of the multiverse may exist as computer simulations (see the argument below) or of course the multiverse may exist as a combination of two or more of the above.

The multiverse is by no means universally popular. As Bob Nichol says, it gives the impression that scientists have 'given up trying to find a theory of everything and simply put the details of our universe down to the shake of a dice'. At the most basic level, the many-worlds interpretation of quantum mechanics appears to violate Occam's razor; never mind not multiplying entities needlessly, this multiplies *everything* and multiplies it *infinitely*. But if parallel universes are unsettling, there is an idea doing the rounds which is positively disturbing.

Nick Bostrom is a philosopher at Oxford University and his simulation argument has become quite famous among scientists, philosophers and lay people alike, partly because of its unsettling weirdness, partly because of the fact that it bears an uncanny resemblance to the plot of the popular sci-fi film *The Matrix* and mostly perhaps because although it sounds crazy, it has so far proved impossible to wholly dismiss out of hand.

The simulation argument goes like this. First of all, we need to make an assumption, namely that one day it will be possible to simulate consciousness inside a computer. If this assumption is incorrect, if consciousness turns out to be something impossible to simulate, then the simulation argument fails immediately. But if it does indeed turn out to be possible to simulate a thinking self-aware mind in a machine, we can then go on to make some more, lesser, assumptions.

First of all, we can assume that one day this will indeed be done. Bostrom points out that computer power is roughly doubling every 18 months or so. In a few decades or less it should be possible, at least in theory, to build machines with a processing power equivalent to that of the human brain. And we must assume that this will indeed be done. The simulation argument then goes on to suppose that programmers will indeed use some of this processing power to create artificial consciousnesses inside their machines, artificial universes in which they can 'live' and furthermore that they will do this more than once (the effect being multiplied greatly by the fact that many of these simulated beings will run simulations themselves).

Again, this seems at least plausible, as even today there are many instances of computers being used to simulate all sorts of real-world scenarios, from the complex weather systems modelled by forecasters to the massive, multiplayer online games in which avatars and computer-generated 'characters' populate imaginary software worlds.

As I write, a story has appeared in the press about the city of Porcupine. In Porcupine, it seems, there have been street protests concerning the establishment in the city of new offices run by the far-right French political party, the *Front National*. The protests escalated, in early 2007, to the point where anti-Nazi activists clashed with right-wing thugs. This is all just mildly interesting until you realize that Porcupine is not a real place, but part of a simulated world, an online virtual universe called Second Life. Online worlds are not new – I seem to remember a couple around from the mid-1990s, but they were slow and clunky and horrible to use and no one knew or cared about them outside the world of the hardline geek.

Second Life might just be different. The software is impressive, creating a colourful, even elegant world of vivid scenery. Second Life has, its makers claim, more than 2.4 million users. And, most importantly, it is becoming more than a computer game. The Second World uses its own currency, Linden Dollars, which can be exchanged for real US greenbacks. At the moment, the only intelligence in Second Life is that of those two and a half million users – that plus some clever software. But it can be safely assumed that this is very much work in progress. Just as scenery can be simulated, so can the 'personalities' of the avatars and other beings with which you, the user, will come into contact. At the moment, and it is worth stating this over and over again, we have no idea whether anyone will ever be able to build any kind of conscious awareness into a computer program. But let's just assume we can, and furthermore that platforms like Second Life will one day become homes for large numbers of these artificial minds. Computers will then be simulating consciousness millions, billions, trillions of times, on millions of machines.

There are no time limits on any of this by the way. These conscious simulations could be built in the 2020s, in the thirtieth century or millions of years hence. It doesn't really matter. And it doesn't matter whether it is we humans doing this, or space aliens living on another planet or even beings living in another parallel universe. All we need to assume is that at some point in the entire history of the Universe, or all the universes, computerized simulations of conscious life are created.

And now, the crux. Because we have assumed that these computerized avatars have been created over and over again, it is overwhelmingly likely, by the sheer and simple weight of statistical probability, that we are living in one of these simulated worlds, a future (and rather more impressive) Second Life, possibly sitting in some adolescent's bedroom, rather than in the one 'real' universe. (Of course, it is just possible that our world *is* the real and original world, but just very, very unlikely.) What then?

Well, everything we think about reality would be wrong. Our Universe, and us within it, would be a fake. Life would be, in effect, a gigantic computer game. Our world would be like a supercharged version of the world of *Doom* or *Grand Theft Auto*, albeit rather more violent. We as individuals would, in a real sense, be reduced to being no more than the playthings of imperfect gods – gods who in our case may well be labouring under the delusion that they themselves are the real McCoy.

It is a profoundly depressing scenario. Is it testable? Well, it is possible to test some of the counter-arguments. One is that is would be impossible to be conscious and not aware that one's situation was not real. This is clearly not the case, because it is possible to dream. When we are dreaming we are in a state of altered consciousness and do not usually know we are dreaming. It is certainly possible to dream while thinking that we are awake. It is also possible to ingest, inhale or inject certain chemicals which have a profound affect on your perception of reality. Being conscious is no guarantee of a grip on what is real.

Mathematically, I am told that the simulation argument stacks up – provided the assumptions it makes about computer technology are correct. Bostrom himself does not *assume* that computers will be able to be conscious; his argument is merely predicated on what would happen if they could, which seems at least plausible. He also allows for the possibilities that no civilization has ever survived the transition to technological maturity, or that no civilization will ever develop an interest in creating a simulated reality.

There are other possible ways to test the argument. Presumably it would be quite hard to make a simulated universe that was completely perfect and consistent in all its parameters – harder at least than making one with a few bricks missing, a few holes around the edges. Do we live in such a universe? Well, as it happens there are some rather big bricks missing in our best models of the Universe. It has so far proved impossible, for example, to reconcile quantum physics and relativity.

?

We cannot see these parallel universes (although it is not impossible to imagine that one day we may be able to build machines to detect them). And we have no evidence, save some mathematical trickery, that we live in a simulated world. The problem with thinking about alternative realities is a paucity of evidence.

Slightly less disquieting perhaps than the idea that we are all the figment of some machine's imagination is the possibility that the fundamental basis of reality is information. John Archibald Wheeler, a quantum physicist, has written that 'what we call reality arises in the last analysis from the posing of yes–no questions'. Can we think of the Universe as a huge cosmic computer, with the ultimate, most fundamental particle being not the quark or the string but the bit of information? After all, everything we know about the Universe is distilled from observation and theory, and that means information.

?

Andrew Liddle, a British physicist, is one of the world's leading cosmological thinkers. He says 'big' questions can be divided into three categories. Firstly, in Category A, are those about which no one has the slightest idea what the answer is, nor how to go about finding one. Category B comprises those questions about which 'there are some theoretical ideas about which the answer might be, but no observational evidence or realistic hope of obtaining any such evidence'. And, finally, Category C: 'there are some ideas about what the answer might be, and some hope of finding observational evidence for or against any given ideas'.

'One answer to the question of existence', Liddle says, 'is the anthropic principle'. In essence the anthropic principle states that things have to be the way they are because if they were not we would not be around to observe them and to ask questions about them. 'The fact that we are here to ask questions and make observations necessarily implies that things have to exist. In this view, it would have been perfectly possible for nothing to have existed, but it is simply impossible for anyone to be able to ask the question unless things do indeed exist'.

Most of the 'big questions' of cosmology and reality, whether or not they stray into the realm of metaphysics, can be put into one of Liddle's three categories. Take the possible arbitrariness of the laws of physics, one of the biggest problems for science. Do the laws of physics have to be as they are?

'It is now widely believed that there is some arbitrariness to the laws of nature we observe', Liddle says. 'For example, there is no reason why gravity isn't quite as strong as it is.' Part of this arbitrariness may come about because the laws of physics may not be immutable and universal. The gravitational constant may have been different billions of years ago from the value it is today, for example. Our visible universe is limited in size to the distance light has been able to travel since the Big Bang.

Some cosmologists believe that the physical laws vary in the very large-scale Universe. At a distance of a googol light years away, for instance (far beyond our observable horizon), the speed of light may be different. Or elementary particles might have different properties. The current front runners in the field of candidates for 'theory of everything' are string theory and its updated cousin M-Theory. In some interpretations of string theory it is predicted that the physical laws will be different in different places in space and time. But the laws may not be enough. Scratch deeper into reality and we are forced to confront the question of what it is that underpins the laws.

Indeed, *is* there a deep underlying logic there at all? One possibility is that we are mistaken in supposing that there is order and not chaos at the bottom of it all. Humans seem to have an inbuilt need to impose mathematical order, symmetry and cause-and-effect relationships on a natural world that often may not work in that way at all. The question of underlying logic probably falls somewhere between Andrew Liddle's categories B and C. Superstring theory is considered quite a good candidate for an ultimate theory of everything, from which all physical laws can be derived. It may thus be considered to be the 'underlying logic' for all these laws.

We are really not much closer to answering the question of why there is anything there at all than were the Ancient Greeks. The question still falls, as Andrew Liddle says, firmly into Category A – 'no idea at all, and no idea how to go about finding an answer'.

The grand swoop of lights we see on a clear night sky is impressive enough; knowing that all those twinkling lights are not only a mere tiny fraction of all the stars out there but in addition that all the stars together possibly form just a small part of what *is*, is humbling beyond belief. The question of reality and what it is has been asked by philosophers and theologians for centuries. Now the baton has passed to science. It remains to be seen whether experiment and observation will end up enlightening us any more than the elegant reasoning of old.

Index

THE SCIENCE OF THE HITCHHIKER'S GUIDE TO THE GALAXY
by Michael Hanlon
MACMILLAN; ISBN 1–4039–4577–2; £16.99/$24.95; HARDCOVER
ISBN: 0–230–00890–9; £8.99/$14.95; PAPERBACK

"Adopting Adams' witty, punchy style, Hanlon's guide is a fun and vivid read. The science twinkles a little more than usual in such a zany setting... he tackles a wide range of cutting-edge topics with depth and authority." *Nature*

"Hanlon's book probes the possibilities inside the fiction with wit and scientist humour – not that you have to be a boffin to enjoy these ruminations, merely curious, as the late Adams himself clearly was." *The Herald*

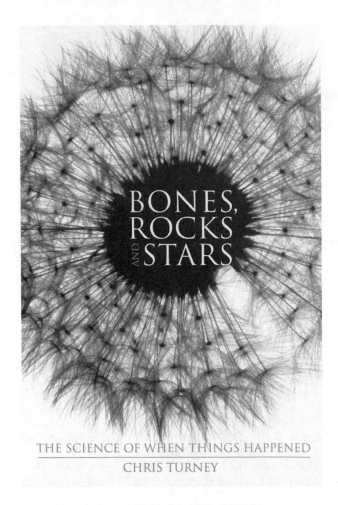

CLIMATE CHANGE BEGINS AT HOME
LIFE ON THE TWO-WAY STREET OF GLOBAL WARMING
by Dave Reay
MACMILLAN; ISBN: 1–4039–4578–0; £16.99/$24.95; HARDCOVER;
ISBN: 978–0–230–00754–3; £8.99/$14.95; PAPERBACK

"Dave Reay has succeeded where so many scientists, academics and environmentalists have failed – in bringing climate change down to the level of the ordinary family. If you're not convinced about climate change, this book will change your mind. It may even change your life." **Mark Lynas**, author of *High Tide*

"How can David Reay be this wise, and still so funny? If you want to get to grips with your own CO_2 emissions – from air freighted grapes to the family runaround – this Edinburgh boffin has written a brilliant, incredibly motivating book. Read it and see." **Nicola Baird**, *Friends of the Earth*

SPACE ON EARTH
SAVING OUR WORLD BY SEEKING OTHERS
by Charles Cockell
MACMILLAN; ISBN 0–230–00752–X/978–0–230–00752–9;
£16.99/US$24.95

"Compelling and well-written – easily accessible to the layman and the expert. If there is to be a bright future ahead of us, the goals set down by Cockell will surely be at the heart of it." *Astronomy Now*

"Cockell's fascinating, impassioned book could convert even the most skeptical – infectious." *Publishers Weekly*

order now from www.macmillanscience.com